Commissioning, Qualification and Validation

Priscilla Browne

ISBN-13: 978-1547091256

ISBN-10: 1547091258

Contents

4. Design Requirements

5. Risk Management

6. Validation Planning

7. Clean Utilities

8. Equipment Validation

9. Process Validation

10. Test Method Validation

11. Supplier Validation

12. Summary of Good Manufacturing Practices (GMP)

Glossary

CHAPTER 1

Introduction to Commissioning and Qualification (C&Q)

Introduction

The term "Commissioning and Qualification (C&Q) refers to Engineering activities primarily related to:

➤ Factory Acceptance Testing (FAT)
➤ Site Acceptance Testing (SAT)
➤ Design Qualification (DQ)
➤ Installation Verification (IV)
➤ Functional Verification (FV)

The below activities are classified as formal Validation deliverables, however, C&Q actions may support the delivery of such requirements.

➤ Installation Qualification (IQ)
➤ Operational Qualification (OQ)
➤ Performance Qualification (PQ)

FAT: Factory Acceptance testing is an engineering activity that is completed at the equipment manufacturer's site or factory. The intention of FAT is to ensure the build and manufacture of the equipment or system shall fulfil the requirements outlined in the User Requirements Specification (URS). FAT provides the opportunities to ensure any critical omissions or errors are addressed in advance of the release and shipping of the equipment. It can be also an important learning opportunity that can ensure the formal qualification tests completed in equipment and process validation are properly designed and considered.

SAT: Site Acceptance testing is also an engineering activity conducted when equipment arrives onsite. Depending on the company, SAT can be completed by the vendor or by the purchasing company. It consists of a series of installation and operational checks to ensure the equipment has not suffered any damage or deterioration between the disassembly, crating, shipping and delivery of the equipment.

DQ: A formal verification that the proposed design of the facility, utility, equipment, system or process is suitable for the intended purpose and meets all requirements.

In practical terms, Design Qualification involves the review of design documentation providing by the vendor or manufacturer. Several stake holders typically have input into design qualification. As many systems or equipment are cross-functional in nature and require automation, validation, quality, engineering and regulatory expertise.

FV: Functional Verification involves verifying that the design conforms to the specification and is physically verified through testing. It should provide documented evidence that the system is Functioning in accordance with the design specifications, User Requirements and engineering drawings

IQ: The documented verification/evidence that the equipment or system is installed as per the manufacturer's recommendations and meets all user requirements for the intended use of the system.

OQ: The documented verification and evidence that the equipment or system operates and functions as intended over the entire operational range of the equipment/system.

PQ: Documented verification and evidence that the equipment or systems performs consistently and is capable of meeting predetermined specifications.

➤ All project activities should be controlled and documented in a change control under a change management program
➤ Facilities and utilities should be qualified in advance of equipment and process validation
➤ Facilities, utilities, equipment and systems should be prospectively qualified
➤ Risk Assessments should be initiated at the beginning of the project and should be living documents.
➤ Identify Critical to Quality attributes and Critical Process Parameters early on or according to in-house procedures.
➤ All qualification work should be executed according to pre-approved protocols. Protocols should have adequate approvals relevant to the work and the risk of the work. Cross-functional teams should review the scope and extent of qualifications in order to ensure all critical aspects are covered and adequately verified.

C&Q Input documents

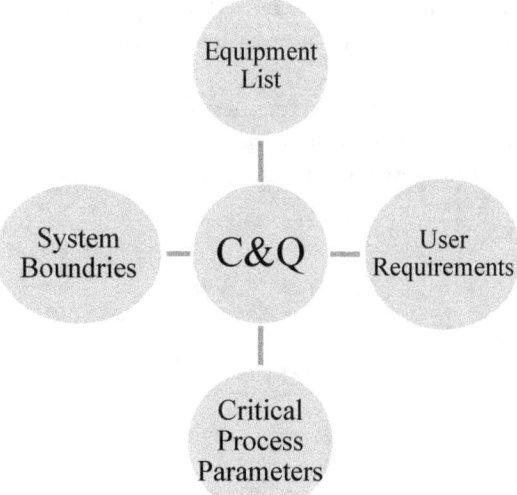

Figure 1: Inputs to C&Q

The detailed list of all equipment and systems in the project scope) is the primary input document for C&Q. The more systems involved, the more documents and activities need to be completed. Complexity of equipment and systems is also a factor. Lists should include both new and modified equipment/systems. This list feeds into key deliverables, resourcing, timelines and implementation dates.

A clear understanding of Systems Boundaries should also be defined at the beginning of a project. The definition of each system / equipment boundaries defines which components on a piping and instrumentation diagram (P&ID), or airflow and instrumentation diagram (A&ID) for heating, ventilation, air conditioning (HVAC) or Process Control architecture belong to which system. This avoids any potential gaps in commissioning and qualification and also ensures responsibilities are crystal clear.

C&Q Approaches

Commissioning activities cover all aspects of both GxP and non-GxP equipment, systems, and processes while the qualification activity is concerned with the GxP aspects. Both demonstrate that the equipment / systems meet the design and contractual requirements, that they function as required for the purposes intended according to the user requirements.

Non GxP Commissioning activities should be properly documented in accordance with Good Engineering practice and according to a defined procedure or guidance document. The following guiding principles should be considered:

➢ Factory acceptance test (FAT): depending on the size, level of customisation, complexity and capital cost of equipment or a new process FAT testing may be required. Generally, small, inexpensive

➢ Minimal commissioning activities are required for off-the-shelf / catalogue items (i.e., verification of conformity with purchase order upon receipt and installation)

➢ More extensive commissioning may be required for complex or custom-designed systems. For example a complex system or process may involve the integration of several pieces of equipment. Integration testing may be necessary to prove the functionality of same.

The three main approaches are to Commissioning:

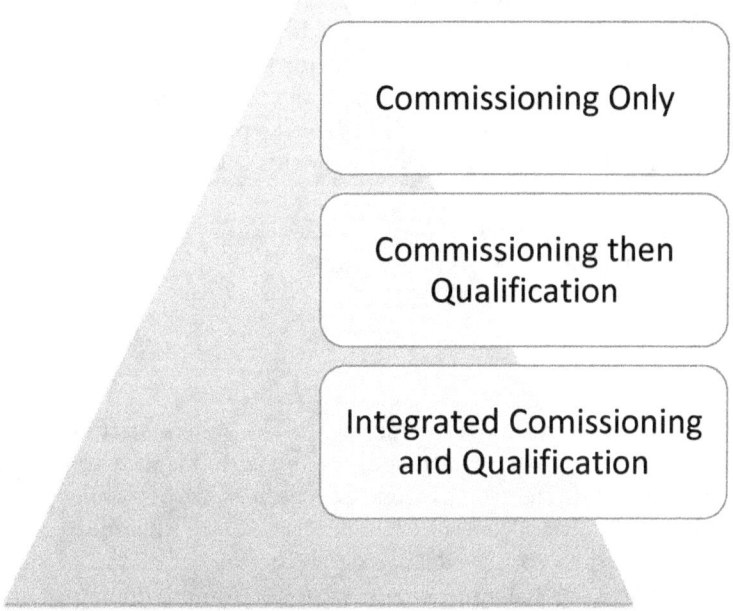

Figure 2: Approaches to C&Q

1) **Commissioning only:** non GxP impacting systems can have a limited level of testing and verification. A formal assessment should be completed and documented to determine if a system is non GxP. Commissioning is therefore a limited engineering activity that should demonstrate that the equipment / systems meet user, design and vendor requirements.

2) **Commissioning then Qualification:** based on the traditional V model known for its "V" or funnel shape. The start of the V model represents gathering user requirements and specifications. Once these are defined and documented in design documents, the inputs can be verified against tangible outputs.

   A "Leveraging" approach where the C&Q plans identify what may be used or leveraged from commissioning to satisfy qualification requirements. Any leveraging should be done with due regard to potential risks and awareness of the criticality of requirements. This avoids repeat testing during the qualification activities, since many of the test required for commissioning and qualification are the same. The conditions allowing leveraging commissioning (e.g. FAT, SAT) tests should be defined in the qualification strategy document such as a qualification or project plan.

3) Integrated Commissioning and Qualification: where the C&Q strategy details the common tests required for commissioning and qualification at the same time and defines the qualification acceptance criteria of the systems, saving time and effort in the project. Based on project size, complexity, skills of team and contractor, the Project Manager jointly with the Quality representative, shall assess which approach is the most appropriate one. Vendor test documentation may be used as part of C&Q, provided that an evaluation of the vendor has been performed. The method of assessment (e.g., questionnaire, audit) shall depend on the criticality, complexity, and impact of the system being provided by the vendor, as well as previous experience with this vendor. Vendor assessments should consider:

> ➢ Does the vendor have a quality management system?
> ➢ The vendor is capable of performing the work
> ➢ The vendor has procedures in place

System Start-Up

After the initial commissioning (siting and utility tie-in) the system start up should be done in a controlled manner. The purpose of a controlled system start up is primarily to prevent injury or damage. As the equipment, system or process may be new to personnel, a well thought out approach to "turning the switch" should be in place. Once the initial system start up is completed, the focus can then shift to debugging and testing of all ancillary systems (utilities) calibration and pre-SAT functional and operational tests. Commissioning is separate to formal equipment validation. It must also be distinguished from formal PQ activities which typically uses product where product outputs are verified to be within pre-defined limits.

C&Q Closure

Good Engineering Practice (GEP) requires a minimum standard of documentation to be provided that is accurate and adequately detailed enough. This should be evident from the beginning of a project (e.g. design documentation). Vendors or contractors must be suitably experienced and capable of delivering accurate technical documentation that is controlled and managed accordingly.

There are typically many individual reports that developed during a C&Q project. Reports are often based on the systems or individual equipment elements. However a final Commissioning Report should be issued to summarize the completion of all activities defined in the Project plan or Commissioning Plan. Reports should cover the following key points.

> ➢ List of commissioning deliverables along with protocol and report references and approval dates

> ➢ Summary list or index of design documentation

> ➢ List and summary of critical non-conformances or deviations

> ➢ Assurance that all open items, non-conformances and deviations are closed and no further action is required.

Decommissioning

Decommissioning is a process to be performed when system / equipment (GxP or non- GxP) will no longer be used. In some upgrade projects, it is part of the project scope under Engineering's responsibility and shall be completed before turning over the modified or new system to the End User.

Decommissioning is the end of the lifecycle of a facility, utility, or equipment when it is deemed to be obsolete, irreparable, is being phased out of use, etc. Decommissioning shall normally include removing the equipment from the facility. There may be occasions when the equipment cannot be removed in a timely manner and, in those cases, steps must be identified in the decommissioning plan to ensure that the equipment is not used (e.g., tags, locks, disconnection from utilities). The minimum requirements to ensure that facilities, utilities, and equipment at the end of their lifecycle are permanently and safely decommissioned are:

> ➢ For GxP systems / equipment, decommissioning must be performed under a site change control. For non-GxP systems / equipment, decommissioning is performed using the appropriate site process (e.g. engineering change management, project plans, site change control)

> ➢ It must satisfy global, site, and regulatory agencies' requirements, including ensuring that the qualified state was maintained until the end of its operation

> ➢ For large decommissioning projects involving numerous systems and equipment, the Team shall create an overall decommissioning plan and use it to facilitate the coordination and sequencing of activities. It will be managed under the same rules as C&Q documentation, including:

> ➢ Use of system boundaries when developing decommissioning protocols or checklists used to record task completion, including documentation supporting the completion

> ➢ Development of a summary report to document the completion of all planned activities.

CHAPTER 2

Facilities

Introduction

Facilities and utilities qualifications are typically pre-requisites to the validation of manufacturing equipment and systems. Much of the activity that deals with establishing a facility or building that is *fit for purpose* is managed under the broad heading of commissioning and qualification (C&Q). The terms C&Q are often used interchangeably and in practice some overlap in activity is expected. Commissioning can be defined as the planned, documented, and managed engineering approach to the start-up and handover of facilities, systems, and equipment to the end-user. It must deliver a safe and functional environment that meets the pre-defined design and user requirements.

In strict terms, qualification is more concerned with the confirmation and documentation showing that equipment or systems are properly installed and functional. Qualification forms part of validation, but the individual qualification steps do not equal a validated process. The establishment of a user requirements specification (URS) and detailed design specifications ensure that the building or facility will meet end users' needs and that it is fit for the intended purpose.

It also provides a level of protection to the contracting company responsible for the project or facility construction. Post-URS approval requires an approved Design Qualification (DQ). This provides verification and a documented record that the proposed design is suitable for the intended purpose. Further verification including IQ/OP/PQ should be applied as required based on the system impact and criticality of facilities/utilities.

Risk and Impact Assessment

A risk-based qualification process should assess the potential of a system to impact the product quality. The boundaries of any system (HVAC, compressed air supply etc.) should be identified in order to help establish the scope of any system and determine if it has a direct, indirect or no impact on product quality.

Direct Impact: a system that can directly impact product quality.

Indirect Impact: where a system is not expected to directly impact the product quality but supports or is ancillary to a direct impact system.

No Impact: a system that does not directly impact product quality and does not support a direct impact system.

For a HVAC System supplying a classified area, only once the air enters the room must the air quality meet the classified designation. The Critical Quality Attributes (CQAs) are routinely monitored through the Environmental Monitoring Program and the Critical Process Parameters (CPPs) should be monitored through the calibrated and validated Environmental Monitoring System (QBMS). The Direct Impact (level 1) for the HVAC systems are indicated on the boundary diagram shown below.

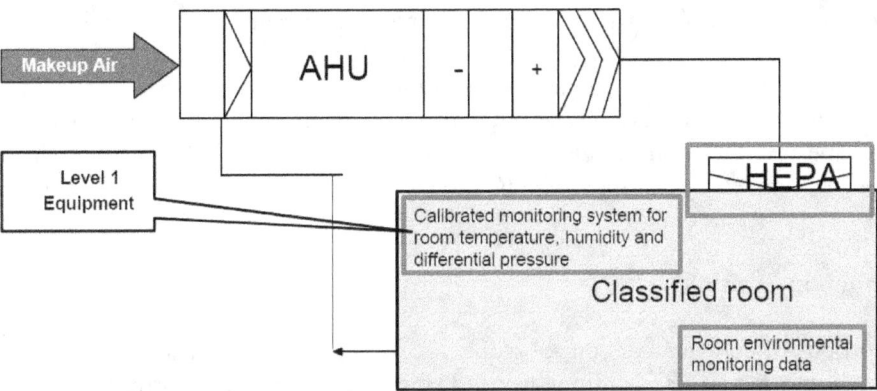

Figure 3: HVAC System Boundary Diagram (Level 1). Each individual system is represented by a green box. Separate qualifications should be performed for each one. The room environmental monitoring system is typically called a Building Management System (BMS). The *calibrated* monitoring system for room temperature, humidity and differential pressure is called a QBMS. Where the "Q" stands for quality indicating the system is used to monitor critical parameters.

Qualification Levels

Qualification levels are often used within companies to classify the criticality of equipment or systems. Level 1 requires the highest level of verification.

Level 1: a system where an **undetected change** in system performance poses a significant risk to the product and product safety. Level 1 systems require the highest degree of qualification and validation. This should include URS/DQ/IQ/ OQ/ PQ.

Level 2: a system where a change that **may be detected** in system performance poses a significant risk to product and product safety. These systems require a level of qualification including IQ, however OQ and PQ testing may not be required. This should be based on the intended use of the system, impact on product quality and overall risk.

Level 3: All other systems.

Typically IQ or equivalent testing is sufficient. **Note:** other requirements or qualifications should be based on risk.

The level of qualification and validation testing required for any system should be based on a risk assessment, examining the criticality of the system and environment. Risk assessments should consider the following points:

- Building design and construction features
- System boundaries and complexity
- Potential product impact
- Environmental controls and monitoring systems
- Potential impact to operator safety
- Type of qualification/validation (e.g. prospective, concurrent, or retrospective)

Controlled-not-classified (CNC) environments, utilities, and facility control systems also require adequate qualification/validation. Again, the impact on product quality should be determined in order to shape any validation. Routine monitoring test locations as well as alert and action levels should be determined in advance of any validation for environmental monitoring or utility systems.

Contamination Control

The philosophy of containment control requires it to be applied across all inputs that make up a facility- equipment, processes, and utilities and so on. Containment is primarily concerned with keeping things in- preventing product or processing agents from egressing into the surrounding atmosphere. Ensuring adequate containment protects personnel who interact with the process, equipment and systems. Aseptic processing often deals with biological agents or compounds that may be harmful to operators or technicians. A secondary concern of containment is protection of the environment. Containment also compliments efforts in contamination prevention. As with Aseptic processing the risk to the patient and product must be at the forefront of activity. Risk based approaches and tools should be used to identify potential risks and put in place adequate controls and mitigations. Any assessment should take into account all the following systems:

- Facility layout
- Drainage Systems
- HVAC requirements
- Location and adequacy of utilities
- Personnel flow and procedures for entering and leaving
- Behavioral requirements of personnel in the clean room
- Flow of materials and products to prevent cross-contamination and mix-ups between products and between dirty and clean or sterile and non-sterile equipment and products
- Design to avoid cross-contamination when manufacturing live biological agents, e.g. local exhaust air HEPA filtration, dedicated air handling units.

Material flow

The design and layout of any manufacturing area should facilitate the effective flow of materials. This is a fundamental requirement no matter what the industry, e.g. medical devices, pharmaceuticals, bio pharmaceuticals and even non-regulated engineering companies that assemble, machine or fabricate products. However, with the manufacture of medicinal products that are required to be sterile imposes a greater level of control and thought. With regard to Aseptic processing facilities material flows do not only require efficient and effective flow of materials, the activity should support the requirements of Aseptic processing while minimizing any risk of contamination. Identifying critical processing zones is step in ensuring the right building design and controls are implemented. Isolators and Aseptic filling require the highest classification with strict environmental controls. Secondary packaging operations such as cartoning are often completed in areas controlled and operated to a lower classification.

Design and layout of facilities should:

 ➢ Maintain microbiological integrity of the identified critical processing zones
 ➢ Prevent or minimise contamination from outside critical processing zones
 ➢ Control the flow of materials by restricting access to trained and authorized personnel

Material Transfer

Material transfer from the outside of cleanrooms to the inside is completed via material air locks or hatches. Material air locks and hatches ensure that there is clear separation between controlled clean areas and less clean areas. Many suppliers provide products that are double bagged. This provides an added level of control when transferring materials. The outer bag can be removed within the air lock thus providing a clean inner product. Material air locks also allow the sanitization of products. Tools and other items must be clean and dirt free.
Controls that prevent personnel from the clean area and less clean area been present in the material air lock at the same time. This can be achieved by training and educating staff on the importance of contamination control. A simple visual check of the air lock to confirm it is vacant can be done in order to avoid mixing of personal from different zones. Decontamination procedures are necessary to ensure materials or tools entering the controlled area are decontaminated.

Material Air lock considerations:
 ➢ Interlocked doors
 ➢ Access control
 ➢ Sanitation/ Cleaning procedure
 ➢ Double or triple bagged products
 ➢ Dedicated trolley for air locks

Disinfection and Cleaning Agents

When materials are been transferred via an air lock, consideration must be given to the status of materials and products. As a rule, no cardboard or unnecessary paper should enter a cleanroom. Wooden pallets are not acceptable as they can carry dirt and microorganisms and wood cannot be sanitized due to its porous nature. Soft fabric cases often used to carry tools should also be avoided as the material can carry dirt and grease. Cleaning and disinfecting agents should be tested and approved prior to their use onsite. The choice of agents should be backed up with studies that demonstrate the effectiveness of disinfectants and cleaning agents.

Gown up Areas

Gowning rooms are designed in order to minimize contamination and facilitate the orderly change over from street clothes to scrubs and/or gowns. Hand washing facilities help reduce the risk of humans carrying unwanted microorganisms into the aseptic processing area. The design of the room should result in clear separation between the less clean side and the clean side. This can be achieved with a step over segregating the two areas.
Other features of gowning rooms should include:

- ➢ Storage lockers for street clothes
- ➢ Gown and Garment storage
- ➢ Body Length mirrors
- ➢ Hand Washing /Drying and disinfection facilities

GMP Zoning

Selecting a suitable classification for a room or manufacturing facility depends on several factors. Firstly, it can be said that sterile products require a more stringent set of criteria than non-sterile products. However, there is an extensive range of products and medical devices that are sterile but are used in different ways and consist of different materials and technology. Some sterile products are single use only and used for short term purposes and then disposed of. Other sterile products are used subcutaneously for longer periods or even require implantation. Therefore, the design of a facility along with its HVAC specification must be appropriate to the product being manufactured. High risk products require greater control. The goal of facilities and HVAC systems is to minimise contamination and the associated risks. Using a "sterile versus non-sterile" rule of thumb is not adequate when classifying a room or facility. Standards including EN ISO 14644-1 and guidelines such as EU cGMP Guidelines EudraLex volume 4 Annex 1 (2008) should be consulted in order to fully understand the requirements of each ISO classification and grade of room.

ISO classifications do not specify room occupancy states but when a designation is applied, the occupancy state must be stated in the relevant documentation or procedure. The most relevant European Guideline (Annex 1 of the EU cGMP Guideline) lists four classification grades and their associated particulate limits in the 'at rest' and 'in operation' conditions. In general, for the sterile and non-sterile products, similar classes are applied, but in non-sterile production the producer could assign their classes, having similar particulate concentration, temperature, pressure etc. but lower air-change rate could be used.

Types of Contamination:

- cross contamination (of a product/material with another product/material)
- non-microbial particulate contamination (non-viable particles)
- biological/microbiological contamination (viable particles/micro-organisms)

Factors Influencing Contamination Cleanliness Levels in the Manufacturing Processes:

- process
- air cleanliness
- personnel hygiene and clothing
- work practices
- material design (material of construction, surface finishes, room finishes, equipment, open system/enclosed system
- utensils, etc.)
- material cleanliness

Room Air Classification (By Limits of Particulate Contamination)

ISO CLASS	FDA	cCMP	Permissible particle number in 1 m3					
			0,1 μm	0,2 μm	0,3 μm	0,5 μm	1 μm	5 μm
1			10	2				
2			100	24	10	4		
3	1		1,000	237	102	35	8	
4	10		10,000	2,370	1,020	352	83	
5	100	A	100,000	23,700	10,200	3,520	832	29
6	1,000	B	1,000,000	237,000	102,000	35,200	8,320	293
7	10,000	C				352,000	83,200	2,930
8	100,000	D				3,520,000	832,000	29,300
9						35,200,000	8,320,000	293,000

Figure 4: Table showing ISO classes and EudraLex Grades A-D.

	Maximum permitted number of particles per m³ equal to or greater than the tabulated size			
	At rest		In operation	
Grade	0.5 µm	5.0 µm	0.5 µm	5.0 µm
A	3 520	20	3 520	20
B	3 520	29	352 000	2 900
C	352 000	2 900	3 520 000	29 000
D	3 520 000	29 000	Not defined	Not defined

Figure 5: maximum permitted airborne particle concentration for each grade. Showing both "at rest" and "in operation" conditions. (EU V4 Annex 1). The EU guidance given for the maximum permitted number of particles in the "at rest" column corresponds approximately to the ISO classifications.

Room Air Classification (By Limits of Microbial Contamination)

The HVAC systems help maintain the viable (microbial) limits within a specific area. These limits are defined in Annex 1 of the EU GMP Guide as shown below.

	Recommended limits for microbial contamination (a)			
Grade	air sample cfu/m³	settle plates (diameter 90 mm) cfu/4 hours (b)	contact plates (diameter 55 mm) cfu/plate	glove print 5 fingers cfu/glove
A	< 1	< 1	< 1	< 1
B	10	5	5	5
C	100	50	25	-
D	200	100	50	-

Figure 6: Recommended limits for microbial contamination

Environmental Grade A (Aseptic)

Grade A is reserved for critical processes in manufacturing sterile products, product components or product contact. This is generally achieved using isolator technology which maintains a barrier to the background environment or surrounding room.

Grade A Operations include:

> ➢ Aseptic processing of sterile ingredients
> ➢ Filling of sterile products not for terminal sterilisation
> ➢ Stopper insertion
> ➢ Crimp Capping

Environmental Grade B

Grade B is used for supportive work for aseptic processing corresponds to ISO 14644 (Part 1) Class 5 ("at rest") and Class 7 (when "in operation"). Grade B areas typically serve as the background environment of Grade A areas for aseptic processing.

Environmental Grade C

Suitable for non-critical processing steps, Grade C corresponds to ISO 14644 Part 1 Class 7 ("at rest") and Class 8 ("in operation"). Grade C operations include:

- ➢ Clean side of material air locks and gowning rooms
- ➢ Filling of products that are to be terminally sterilised

Environmental Grade D

Grade D at least corresponds to ISO 14644 Part 1 Class 8 ("at rest" / no definition for "in operation").

- ➢ Clean section of material air locks and final compartments of gowning rooms
- ➢ Dispensing of raw materials and excipients and preparation of solutions for sterile products to be sterile filtered and terminally sterilised
- ➢ Background environment for transfer and crimp capping of stoppered containers with sterile products

Compliance Tests for GMP Zones

Test	Requirements
Particle count test	Test covers verification of cleanliness. Dust particle counts to be carried out and result printed. The number of readings and positions of tests should be defined in accordance with ISO 14644-1 Annex B5
Air pressure difference	This test is used to verify non cross-contamination. Log of pressure differential readings to be produced or critical plants should be logged daily, preferably continuously. A 15 Pa pressure differential between different zones is recommended. Refer to ISO 14644-3 Annex B5
Airflow volume	To verify air change rates. Airflow readings for supply air and return air grilles to be measured and air change rates to be calculated. Refer to ISO 14644-3 Annex B13
Airflow velocity	To verify unidirectional flow or containment conditions. Air velocities for containment systems and unidirectional flow protection systems to be measured. Refer to ISO 14644-3 Annex

	B4
Filter leakage tests	To verify filter integrity. Filter penetration tests to be carried out by a competent person to demonstrate filter media, filter seal and filter frame integrity. Only required on HEPA filters. Refer to ISO 14644-3 Annex B6
Containment leakage	To verify absence of cross-contamination. Demonstrate that contaminant is maintained within a room by means of: • airflow direction smoke tests • room air pressures. Refer to ISO 14644-3 Annex B4
Recovery	To verify clean-up time. Test to establish time that a cleanroom takes to recover from a contaminated condition to the specified cleanroom condition. Should not take more than 15 minutes. Refer to ISO 14644-3 Annex B13
Airflow visualisation	To verify required airflow patterns. Tests to demonstrate air flows: • from clean to dirty areas • do not cause cross-contamination • uniformly from unidirectional airflow units Demonstrated by actual or video-taped smoke tests. Refer to ISO 14644-3 Annex B7

Clean Room Design Considerations

Air handling unit (AHU) -Air Intake Quality

Seasonal Variations

All locations on earth except latitudes near the equator experience seasonal temperature changes. The changes are a consequence of the earth's orbital motion about the sun, coupled with the tilt of earth's axis of rotation with respect to its orbital plane. Design criteria should be based on published temperature data. The HVAC system design should consider the following:

Standard Operating Conditions: These are climatic conditions against which the systems must be designed to operate, control, and maintain required conditions. (These may be based on published data, which are only exceeded 2.5% or 1% of the time).

Extreme Operating Conditions: These are climatic conditions against which the systems must be designed to operate, without manual intervention, and without damage to the systems or the facility. Based on product / process risk assessments, extreme or standard conditions shall be used for HVAC design for dedicated areas.

Location

Based on the building layout, foot-print and design intent, a suitable and adequate space must be identified for HVAC location. This must include provision of chilled water, heating systems, ducts and drainage. HVAC plants must be accommodated in designated HVAC plant rooms or interstitial areas.

Air Intake

During the design phase, the air intake locations should be selected that ensure air is in the best environmental condition. The below considerations help to achieve a strong starting point:

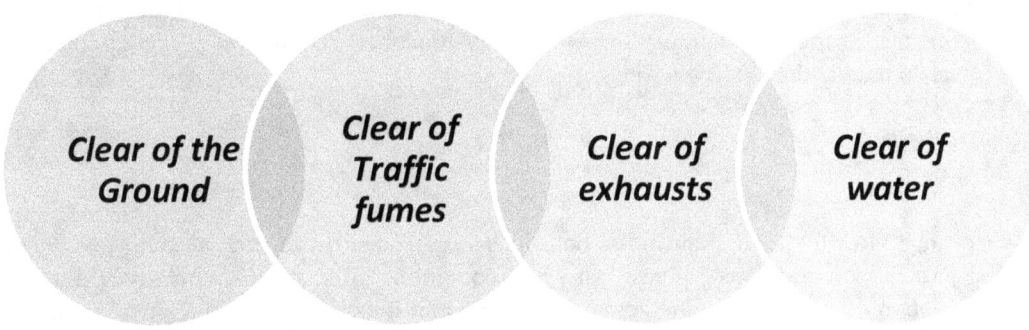

Figure 7: Air intake considerations

Thermal Load

Thermal load can be defined as the amount of heat energy to be removed from an inner environment by equipment (HVAC) used to maintain that environment at the design temperature when worst case external temperature(s) are being experienced. The thermal load requirement should be calculated for the following:

> ➤ Max summer conditions
> ➤ Minimum winter conditions
> ➤ High rainfall
> ➤ Standard operation
> ➤ Extreme operating conditions

Room Recovery Time

Room recovery time to return to the required pressure differential and cleanliness Specification should be minimized.

Dust, Vapour, or Fume Control

Highlight areas requiring dust, vapour, gas and/or fume control on the room data sheet. These areas must be controlled to remove the possibility of product contamination and to ensure the safety of the operator and environment. Areas requiring 100% fresh air or extraction to atmosphere may require greater airflow or other measures within the room to maintain environmental conditions.

In order to meet the appropriate level of cleanliness, HVAC systems require sufficient filtration to provide "clean" air to prevent contamination of the product. Pre-filters and main filters are normally suitable for most operations; however, HEPA filters are required to prevent particulate or microbial contamination for higher-classification areas

Air Change Rates

The air change rates for each room must be calculated to be sufficient for cleanup to achieve specified particulate conditions "at rest" in static conditions after a maximum of 20 minutes from completion of operations. The actual air change rate must be chosen to satisfy the most stringent requirement including GMP, GLP, heat gain, ventilation requirements and/or occupancy, including an appropriate safety factor.

The air change rate must be optimized for energy savings; however, specific attention must be paid to air locks where a greater air change rate must be applied. Air changes can be reduced (e.g., setback modes) in some circumstances ("at rest" mode, with no production activity and no personnel interventions).

Room Environmental Conditions

Other environmental conditions to be controlled, such as temperature and relative humidity, depend on the product and nature of the operations carried out in those areas. These parameters should not interfere with the defined cleanliness standard.

Temperature Requirement

The normal operating temperature requirement for each classification .Temperature and humidity must be appropriate to the product and process Consideration should be made for specific product and process requirements.

Humidity Requirement

The normal operating humidity requirement

Particulate Levels

Particulate levels are specifically defined for each room classification "at rest" and "in operation". The levels are controlled though air filtration, facility design, gowning requirements, and decontamination

Room Exhaust

Where there is a risk of active compounds being present in extracted air, filters should be fitted, preferably at the room, to prevent contamination of ductwork and the environment. The filters must be selected based on the particle size distribution of the products to be handled.

HVAC system design

The HVAC system must be appropriately selected using the specific design requirements as outlined above. The system must be able to provide clean, conditioned air to the specified areas to meet all of the quality requirements. The most important precursor to HVAC design is the comprehensive definition of the function and performance required followed by the selection of an appropriate system. A poor selection can lead to unnecessarily high-energy consumption, and operational deficiencies. HVAC systems can be divided into two main types:

All-air systems rely on the movement of large quantities of air through a central air handling unit to control room conditions, as well as provide for ventilation requirements.

They have the advantage of being relatively simple with most of the unit situated in one location; however, they are very space consuming. All-air systems tend to be relatively inflexible and not ideal for areas that are likely to need environmental alteration on a regular basis.

These HVAC systems are used for areas that have a lot of small zones, each with slightly different thermal loads but which requires constant ventilation. These systems can have poor energy efficiency if a lot of reheat is required. These are typically used in large manufacturing areas, and laboratories with many small rooms.

Dust Extraction and Collection

It is essential to capture dust as close as possible to the point of generation without affecting the process. In most cases dust capture should be within 100 mm from the point of release. Air velocity is the key parameter in dust capture.
Pharmaceutical and chemical applications have specific collection requirements as any dust build-up in the system is likely to be of a pharmacologically active nature, sensitizing, toxic and/or corrosive. It is vital to maintain transport velocities and minimizes any potential for cross contamination.

A typical system should have a minimum transport velocity of 18 m/s, but this may need to be higher if heavy particles are to be collected. This velocity must be maintained throughout the system to prevent dust from dropping out in the ducts.

The dust collection must be configured with the hazardous nature of the dust in mind. A clearly defined disposal procedure for the collected dust (e.g., bag-in / bag-out system for filter and dust bin) needs to be understood at the design stage. HVAC unit shall meet EN 1886 and EN 13053 requirements.

Fans

Certified performance curves are required to verify correct fan operation. Fans that may be subjected to high temperatures, humidity, corrosive fumes or other hazardous atmospheres should be constructed using non-reactive, non-corrosive, suitable and approved materials (such as epoxy painting). Whenever H2O2 or other disinfection application is planned, material compatibility certificates shall be supplied by the vendor.

Fans must be selected to supply the design volume, taking into account the assumption that filters are half clogged, except for the terminal filter which shall be considered to be fully clogged according to EN 13053. If the terminal filter is HEPA, clogging shall be considered according to EN 1822 and the target volume is 80% of the given maximum clogged specified value.

Filtration

Face-fitting filters shall be used in all cases, as slide-in filter elements never give a good seal. The installation must be such that the airflow pushes the filter against the seal. The face velocity across the filter section shall not exceed 2 m/s. For ventilation and air conditioning applications, two minimum filtration stages are required. For certain applications, return air filtration will be required to contain highly active materials (e.g., viruses or potent compounds). Normally, these filters should be changed from the room side. However, since those filters must be integrity tested, it is recommended to place one filter in the main return duct before the exhaust fan and design return duct network, in order to ensure tightness of the duct between the room and the filter (bag-in / bag-out filter change systems should be provided for BSL-3 areas).In case of live biological agent biocontainment, decontamination up to the filter must be proven. The grade of filter and technical solution must be selected based on the product particle size distribution and occupational exposure band (OEB) level.

HEPA filters and Dehumidification

For most HVAC applications, dehumidification is best achieved by the use of cooling coils. It should be noted that dehumidification is a very high consumer of energy and should only be used if there is a real process need. When areas are not in use, the dehumidifier should be turned off, if possible.

When room humidity must be maintained below 50% during warm weather, an absorption dryer may be necessary unless the room temperature can be increased within specification to compensate.

Normal practice is to optimize size and efficiency of the absorption dryer by first removing as much moisture from the air as possible by cooling. The design of absorption dryers is normally based on a slowly rotating desiccant wheel.

Air is passed through the wheel and dried by the desiccant coating (guidance: lithium chloride especially if the wheel is not used frequently and silica gel if used permanently and with low humidity target). It is not normally necessary to size a dryer to handle the entire air volume. Drying a proportion of air and re-mixing to achieve the desired moisture content is usually sufficient.

Air humidification may be necessary during cold weather when introducing fresh air to spaces that require humidity control. When air humidification is necessary, humidifiers should be selected on the following basis:

> ➤ direct steam injection using steam
> ➤ direct steam injection using self-generative electric or gas steam humidifier.

Clean steam is required, However when industrial steam is used a quality gap assessment should be conducted with Quality team taking into account the regulatory requirements 21 CFR 173.310 for operator safety and to avoid product contamination and mainly boiler feed water treatment must be free of volatile additives such as amines and hydrazines.

Humidifiers should be located before the fan and the final filter which will remove any particulate generated. At least 300 mm clearance should be allowed upstream and 1 m downstream between humidifier manifolds and coils, attenuators etc. (general recommendation to be confirmed through calculation note provided by the vendor). A single manifold or multiple manifolds in parallel may be used to meet the humidification requirements as per manufacturer's recommendations.

Sound Attenuators

Sound attenuators should be provided as necessary, to achieve the specified noise levels within occupied spaces and to minimize external noise nuisance assessment can confirm the necessity to use acoustic media (enveloped in polyester film), that is inert and corrosion-resistant at normal operating conditions. Material quality shall be equivalent to that specified for HVAC unit or ducts.
Sound attenuators should be installed in the air handling unit or ductwork. The use of sound attenuators in the air supply and air return should be based on requirements for fresh air inlet and air exhaust, and according to external noise levels that might need to be maintained at or below the ambient site noise levels.

Dampers

The provision of sufficient dampers is essential for proper control. To minimize noise transmission into the room, these should be mounted as far as possible from the diffuser.

Carefully evaluate the space-by-space pressure control that will be used in the design. Static pressure control via hard balance or dynamic control via air terminal control units are both appropriate. Consideration should be given to the overall project size, the complexity of the facility and the project budget.

Automatic volume controllers are recommended for regulating air volume independently of supply pressure. They can be selected for constant volume, variable volume or dualduct mixing applications. Automatic low-leakage fresh air and exhaust air shutoff dampers are strongly recommended to isolate the HVAC network. Fresh air dampers shall be Class 3 minimum (maximum leakage preventing coil freezing). Whenever fumigation is performed shutoff damper shall ensure Class 4 leakage rate. Where dampers are required to provide modulating control of airflow, they must be selected to provide an appropriate level of control authority. This will normally mean a damper smaller than the duct size.

Heating and cooling

Heating mode: Low pressure hot water (LPHW) is the preferred heating medium for HVAC applications and should be used whenever practicable. Electrical heating should be avoided due to fire risk and should limited to low power coil and in locations where no other energies are available. Hazard operability analysis (HAZOP) must be conducted if electrical heating is being considered. Cooling mode: Chilled water is the preferred cooling medium for HVAC applications and should be used whenever practicable.

The direct expansion of refrigerant in coils is an acceptable method of cooling, particularly on small isolated plants, or where lower temperatures are needed for dehumidification or for cold room. This system, however, does not normally give close control. Direct expansion coils should only be used with extreme care on variable air volume systems (if speed driver available on compressors).

Heating Coils

The face velocity of air across heating coils should not exceed 2 m/s. Coils should be made of material suitable for applicable constraints. Drains shall be located outside the casing of the HVAC unit. Coils shall be removable.

Cooling coils

Cooling coils have been identified as potential sources of microbial contamination; therefore, careful design is required to prevent water carryover and to ensure that drain pans do not retain water. Double tube, non-welded units are recommended. The face velocity of air across cooling coils should not exceed 2 m/s. Where necessary, stainless steel or plastic eliminator blades should be provided to prevent any moisture carryover. Where provided, these must be removable for cleaning.

Ductwork

For most applications galvanized steel ductwork will be the most appropriate form of construction; however, stainless steel or plastic construction may be necessary where there is a higher risk of corrosion due to moisture or fumes (exhaust ducts usually). Where operating pressures above 2,000 Pa are necessary, fully welded construction is recommended. For contained ducts (e.g., exhaust duct before bag-in / bag-out filter), air tightness Class C shall be followed (EN 12237). For BSL-3, fully welded construction should be considered.

Generally ductwork should be constructed to an appropriate local standard, suitable for the maximum design pressure (positive or negative), such as those published by Sheet metal and Air Conditioning Contractors' National Association (SMACNA) in the USA, Building and Engineering Services Association (B&ES) in the UK
Where flexible connections are proposed these must be designed for the same pressure as the ductwork. Solid ducted connections are preferred for final connections to terminal HEPA filter housings. For applications where flexible connections to diffusers are used, these should be no longer than 500 mm and nominally straight.

Special consideration must be given to fume extract ducts where these pass through fire barriers. Using fire dampers should be avoided where the loss of extraction could make a fire situation worse. An alternative design, such as the use of fire-rated ductwork, may be necessary in these cases. A thorough risk assessment must be conducted.

Simple Representation of HVAC system

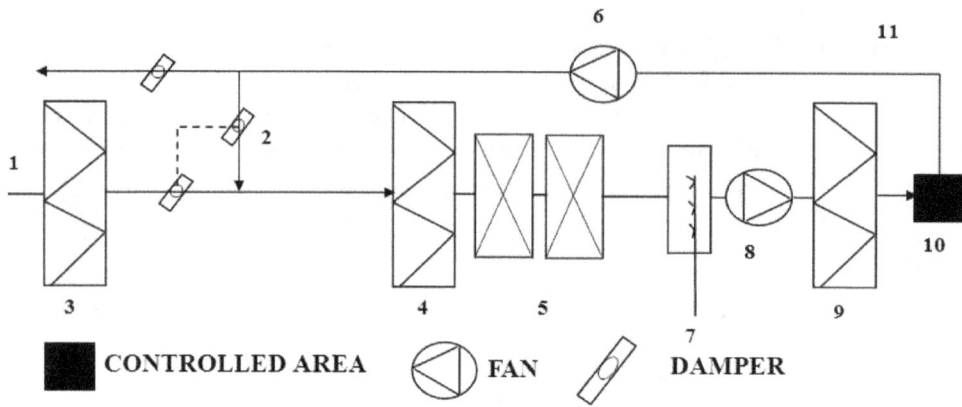

Figure 8: Simple HVAC diagram

Position	Description
1	Fresh Air Intake (°C, %RH, Flow rate)
2	Dampers
3	Filter creating a Differential Pressure
4	Filter creating a Differential Pressure
5	Control Valves for cooling fluid
6	Exhaust Fan
7	Steam Flow Rate
8	Supply Fan
9	Filter creating a Differential Pressure
10	Controlled room/ area
11	Extraction

Parameter	Description
Temperature	The HVAC must be capable of operating over a range of temperatures and accurate to a tolerance. Temperature probes/detectors must be placed at various points to provide feedback and control.
Relative Humidity	Relative humidity must be monitored continuously. Typically, humidity sensors should be effective over an operational range (e.g. 5-95% R.H.) Accuracy should also be no less than ±3%
Air flow/ Air Pressure	Air flow is proportional to the square root of the differential pressure.
Dampers	Dampers are used to control inlet and outlet airflow.
Valves	Ball valves or globe valves control the flow of air. Valves are designed with specific safety features to meet the intended use.(e.g. CLOSED without energy supply-cooling valve, OPEN without energy supply-heating valve.

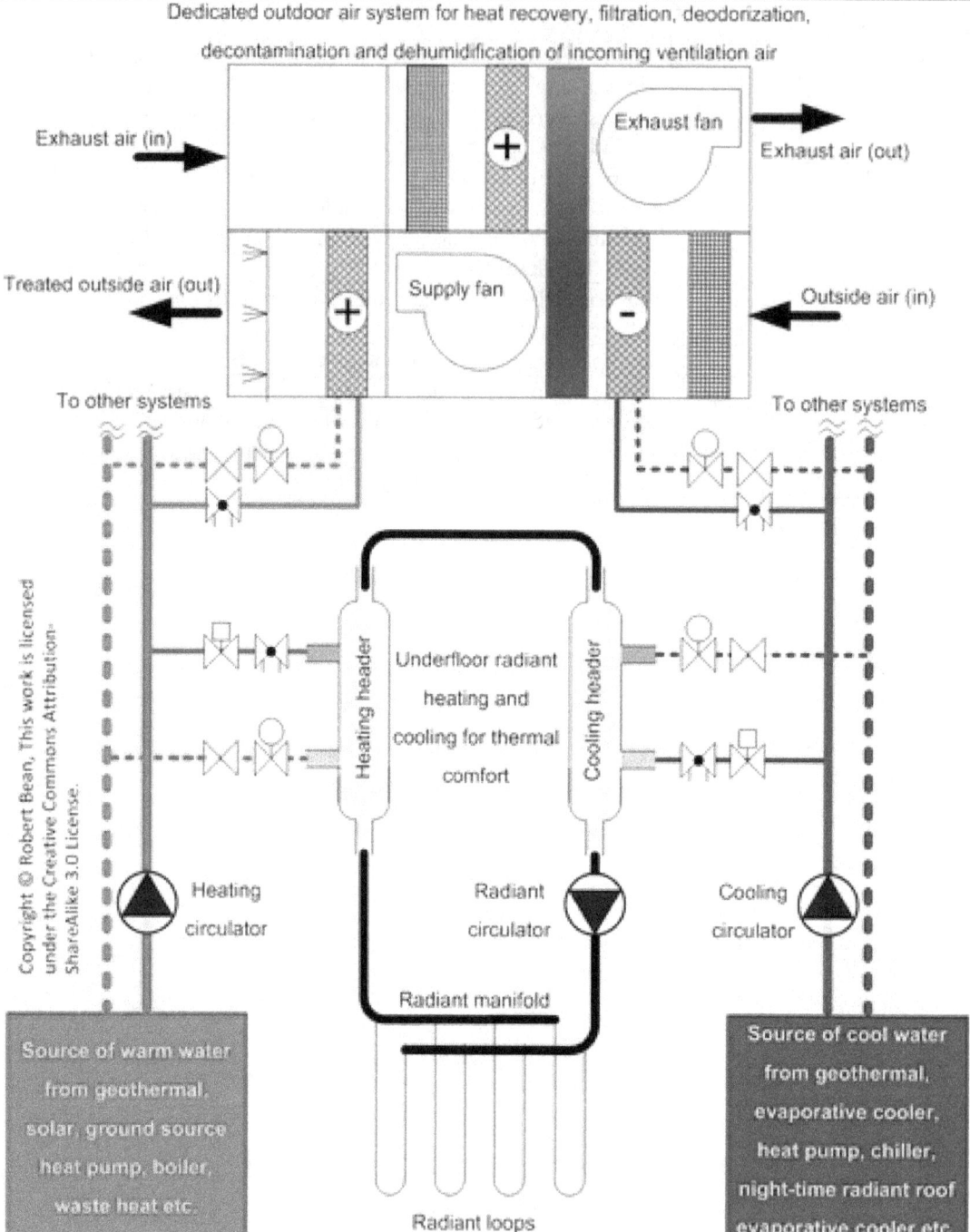

Dedicated outdoor air system for heat recovery, filtration, deodorization, decontamination and dehumidification of incoming ventilation air

Exhaust air (in)

Exhaust fan

Exhaust air (out)

Treated outside air (out)

Supply fan

Outside air (in)

To other systems

To other systems

Heating header

Underfloor radiant heating and cooling for thermal comfort

Cooling header

Heating circulator

Radiant circulator

Cooling circulator

Radiant manifold

Source of warm water from geothermal, solar, ground source heat pump, boiler, waste heat etc.

Source of cool water from geothermal, evaporative cooler, heat pump, chiller, night-time radiant roof evaporative cooler etc.

Radiant loops

Simplified schematic for a radiant based HVAC system.

Some items suh as controls, wiring, expansion and fill assemblies have been deliberately omitted for drawing clarity

Environmental Monitoring

An environmental monitoring program is required for GMP controlled areas. The purpose of such programs is to document, define and describe parameters to be monitored, monitoring frequency and methods. Environmental monitoring is a regulatory requirement. It also demonstrates that the GMP areas are been controlled and are fit for purpose.

Key Requirements of Environmental Monitoring

Identification and classification of environmental areas that require monitoring

Test methods and sampling procedures

Defined testing frequencies

Sample locations based on Risk

Microbial monitoring of personnel

Monitoring of non viable particles

Monitoring of temperature, relative humidity and differential pressures

Defined alert and action levels for each environmental area

Trending of Enviromental data

Change Control

Other parameters such as those controlled by the HVAC system (air changes/hour etc.) should also be verified according to a defined schedule.

Grade A, B and C

> ➢ Viable and non-viable particles monitored under operational conditions
> ➢ Risk-based approach to sampling points and represent high risk/ critical positions

Grade D

> ➤ Non-viable particles must be measured at-rest conditions
> ➤ Viable particles measured under operational conditions

Building Management Systems

A Building Management System (BMS) is an automated control system that is used to manage a building and facilities heating, ventilation and air conditioning, security, fire protection systems and so on. It is made up of many different Input / Output subsystems, controller(s), server(s) and workstation(s) communicating over a control network to control, monitor, alarm and trend equipment. . BMS systems are also referred to as a Facilities Management/Monitoring System (FMS), Energy Management System (EMS), Building Automation System (BAS) or other equivalent.

Environmental Monitoring System (EMS) are automated control systems consisting of Input / Output subsystems, controller(s), server(s) and workstation(s) communicating over a control network to monitor, alarm and trend environmental critical process parameters such as temperature, humidity, differential pressure, conductivity, cooler / refrigerator status amongst others. Suggested classification of BMS based on intended use:

Building Management System (BMS)	
System Classification	GxP
Data Usage	Data not used for GxP impacting decisions. Engineering use only
Monitoring	No Critical process parameters are monitored by the system
Controls	No GxP equipment
System Boundaries	Up to the point of use of the system or equipment
Validation	Not required

Environmental Monitoring System (BMS)	
System Classification	Non GxP
Data Usage	Data may be used to make quality decisions and product release decisions. Data is used to determine compliance.
Monitoring	Critical process parameters are monitored by the system
Controls	Critical alarm limits are controlled
System Boundaries	From the point of use
Validation	Not required

If a BMS system is not intended for use with GxP activities and there is another GxP compliant system to meet product quality / critical parameters, then by a BMS system does not need to be qualified and should be classified as non- GxP system. This is based industry best practices and the intended purpose of a system.

Further reading

ISO 14644-1: International Organisation For Standardisation – Cleanrooms and Associated Controlled Environments. Part 1: Classification of Air Cleanliness.

ISO 14644-3: International Organisation For Standardisation – Cleanrooms and Associated Controlled Environments. Part 3: Test Methods.

ISO 14644-4: International Organisation For Standardisation Cleanrooms and Associated Controlled Environments: Part 4: Design, Construction and Start-Up.

EudraLex, Vol 4, Annex 1: EU Guide to Good Manufacturing Practice (EU GGMP) Governing Medicinal Products for Human and Veterinary Use, Annex 1 – Manufacture of Sterile Medicinal Products.

EN 1822:2009: European Standard For HEPA Filter Classification.

US FDA CFR 211: Code of Federal Regulations Food and Drug Administration Title 21 Part 211 – Current Good Manufacturing Practice for finished pharmaceuticals – Section 211.46 Ventilation, air filtration, air heating and cooling.

ICH Q7: International Conference on Harmonisation - Good Manufacturing Practice Guide for active pharmaceutical ingredients – Section 4.21 and 4.22 – Utilities.

US FDA: Food and Drug Administration - Guidance for Industry "Sterile Drug Products Produced by Aseptic Processing – current Good Manufacturing Practice".

CHAPTER 3

Introduction to Validation

Introduction

The term 'Validation Lifecycle' refers to the entire lifecycle, beginning with the initial requirements of a product or process and identifying CPPs and CQAs. The cycle continues through Commissioning and Qualification (C&Q), PQ and PV, requalification and ending with decommissioning or the end of life of a product line.

Process Validation is defined as "establishing documented evidence which provides a high degree of assurance that a specific process consistently produces a product meeting its predetermined specifications and quality attributes."

Why is Validation required?

There are several factors that require Validation activities within the life science industry. Above all, validation works to ensure patient safety and products that are fit for purpose and reliable time after time. However, regulations provide the legal incentive to validate processes within the medical device and medicinal industry. Validation activity also has secondary effect of fostering consistency in introducing new products and processes across different departments and sites.

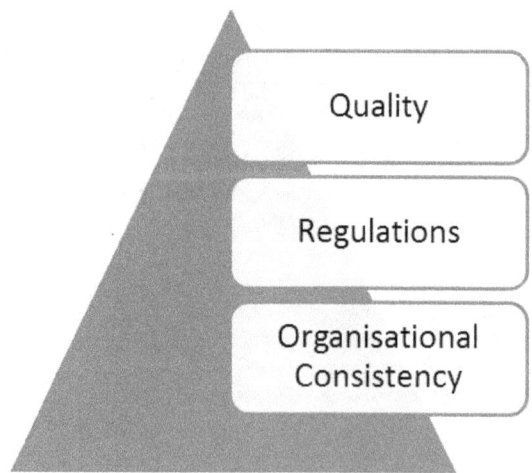

Figure 9: Drivers of Validaiton

In many respects, regulations are the key driver as with any legal requirement, the manufacturer is required to fulfill their statutory obligations. Notified bodies such as the UK, MHRA, US FDA specify rules and guidance in respect of pharmaceutical and medical products.

As the largest economy is the world, the United States is a leading user of regulated products. In turn, many countries throughout the world manufacture products with the intent to supply the

US market. Therefore, manufacturers must meet the regulatory requirements set down by the US FDA. For Medical devices, 21 CFR Part 820 requires Validation to be completed for equipment and processes. For Pharmaceuticals, 21 CFR Part 211 also makes provision for Validation. In Europe, validation for the manufacture of medicinal products is a requirement of EU GMP V4, Medicinal product for human and veterinary use.

Summary of GMP Regulations and Standards

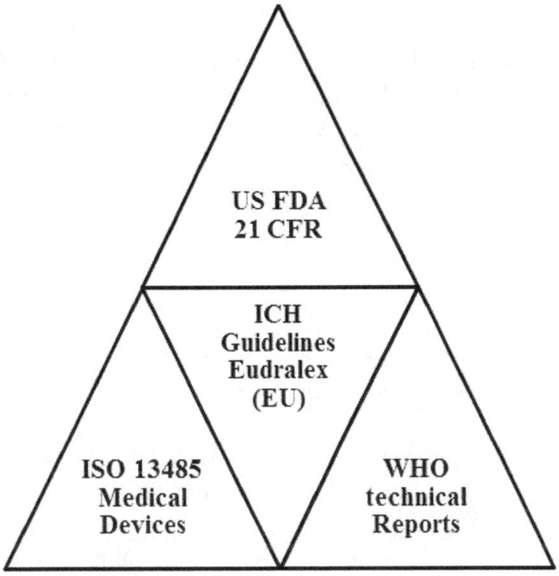

Figure 10: Key regulations and standards

The Four Types of Process Validation

Process validation is a regulatory requirement of Good Manufacturing Practices (GMPs) for both pharmaceuticals (21CFR 211) and medical devices (21 CFR 820).

Prospective validation

Establishing documented evidence in advance of process implementation that a process or system operates as intended. This is the preferred approach and is most common when new products must be validated before commercial manufacturing.

Concurrent validation

Establishing documented evidence that a processes operates as intended, based on information generated during process implementation. Concurrent means that the outputs are performance of the system is monitored at the same time a manufacturing which can include commercial lots.

Retrospective validation

Retrospective validation is used for facilities or processes that have not completed formal Validation. Historical data or a retrospective review can provide the evidence that the process or facility is operated as intended. This type of validation is uncommon.

Revalidation

Revalidation involves the re-execution of validation activities in order to maintain a validated state. This can be a result of substantial changes to Product attributes or specification or changes to the manufacturing process itself. Other reasons a partial or full revalidation may be required involve instances where product quality issues have increased.

Stages of Process Validation

Process validation can be divided into in three stages:

Stage 1 – Process Design: The commercial manufacturing process is defined during this stage based on knowledge gained through development and scale-up activities.

Stage 2 – Process Qualification: During this stage, the process design is evaluated to determine if the process is capable of reproducible commercial manufacturing.

Stage 3 – Continued Process Verification: Ongoing assurance is gained during routine production that the process remains in a state of control.

Before any batch from the process is commercially distributed for use by consumers, a manufacturer should have gained a high degree of assurance in the performance of the manufacturing process such that it will consistently produce APIs and drug products meeting those attributes relating to identity, strength, quality, purity, and potency. The assurance should be obtained from objective information and data from laboratory-, pilot-, and/or commercial scale studies. Information and data should demonstrate that the commercial manufacturing process is capable of consistently producing acceptable quality products within commercial manufacturing conditions. A successful validation program depends upon information and knowledge from product and process development. This knowledge and understanding is the basis for establishing an approach to control of the manufacturing process that result in products with the desired quality attributes. Understanding variation and knowing how to detect and control it is therefore a key element to maintaining robust processes and systems.

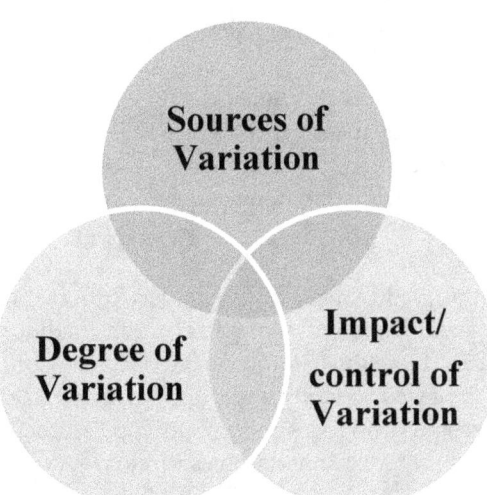

Figure 11: Variation

Manufacturers should, understand the sources of variation, the degree of variation, the impact of variation and how to detect and control variation. Each manufacturer should judge whether it has gained sufficient understanding to provide a high degree of assurance in its manufacturing process to justify commercial distribution. Focusing exclusively on qualification efforts without also understanding the manufacturing process and associated variations may not lead to adequate assurance of quality. After establishing and confirming the process, manufacturers must maintain the process in a state of control over the life of the process, even as materials, equipment, production environment, personnel, and manufacturing procedures change. Manufacturers should use ongoing programs to collect and analyze product and process data to evaluate the state of control of the process. These programs may identify process or product problems or opportunities for process improvements that can be evaluated and implemented through some of the activities described in Stages 1 and 2.

Manufacturers of legacy products can take advantage of the knowledge gained from the original process development and qualification work as well as manufacturing experience to continually improve their processes. Implementation of the recommendations in this guidance for legacy products and processes would likely begin with the activities described in Stage 3.

Stage 1 — Process Design: Process design is the activity of defining the commercial manufacturing process that will be reflected in planned master production and control records.

The goal of this stage is to design a process suitable for routine commercial manufacturing that can consistently deliver a product that meets its quality attributes. Building and Capturing Process Knowledge and Understanding Generally, early process design experiments do not need to be performed under the CGMP conditions required for drugs intended for commercial manufacturing and supply.

Stage 2 (process qualification) and Stage 3 (continued process verification). They should, however, be conducted in accordance with sound scientific methods and principles, including good documentation practices. Decisions and justification of the controls should be sufficiently documented and internally reviewed to verify and preserve their value for use or adaptation later in the lifecycle of the process and product. Although often performed at small-scale laboratories, most viral inactivation and impurity clearance studies cannot be considered early process design experiments.

Viral and impurity clearance studies intended to evaluate and estimate product quality at commercial scale should have a level of quality unit oversight that will ensure that the studies follow sound scientific methods and principles and the conclusions are supported by the data.

Product development activities provide key inputs to the process design stage, such as the intended dosage form, the quality attributes, and a general manufacturing pathway. Process information available from product development activities can be leveraged in the process design stage. The functionality and limitations of commercial manufacturing equipment should be considered in the process design, as well as predicted contributions to variability posed by different component lots, production operators, environmental conditions, and measurement systems in the production setting. Design of Experiment (DOE) studies can help develop process knowledge by revealing relationships, including multivariate interactions, between the variable inputs (e.g., component characteristics or process parameters) and the resulting outputs (e.g., in-process material, intermediates, or the final product).

Risk analysis tools can be used to screen potential variables for DOE studies to minimize the total number of experiments conducted while maximizing knowledge gained.

These activities also provide information that can be used to model or simulate the commercial process. Computer-based or virtual simulations of certain unit operations or dynamics can provide process understanding and help avoid problems at commercial scale. It is important to understand the degree to which models represent the commercial process, including any differences that might exist, as this may have an impact on the relevance of information derived from the models. It is essential that activities and studies resulting in process understanding be documented. Establishing a Strategy for Process Control Process knowledge and understanding is the basis for establishing an approach to process control for each unit operation and the process overall.

Strategies for process control can be designed to reduce input variation, adjust for input variation during manufacturing (and so reduce its impact on the output), or combine both approaches. Process controls address variability to assure quality of the product. Controls can consist of material analysis and equipment monitoring at significant processing. Decisions regarding the type and extent of process controls can be aided by earlier risk assessments, then enhanced and improved as process experience is gained. The planned commercial production and control records, which contain the operational limits and overall strategy for process control, should be carried forward to the next stage for confirmation.

Stage 2 — Process Qualification During the process qualification (PQ) stage of process validation, the process design is evaluated to determine if it is capable of reproducible commercial manufacture.

This stage has two elements: (1) design of the facility and qualification of the equipment and utilities and (2) process performance qualification (PPQ). During Stage 2, CGMP-compliant procedures must be followed. Successful completion of Stage 2 is necessary before commercial distribution. Products manufactured during this stage, if acceptable, can be released for distribution.

Design of a Facility and Qualification of Utilities and Equipment Proper design of a manufacturing facility is required under part 211, subpart C, of the CGMP regulations on Buildings and Facilities. It is essential that activities performed to assure proper facility design and commissioning precede PPQ.

Here, the term qualification refers to activities undertaken to demonstrate that utilities and equipment are suitable for their intended use and perform properly. These activities necessarily precede manufacturing products at the commercial scale. Qualification of utilities and equipment generally includes the following activities:

➤ Selecting utilities and equipment construction materials, operating principles, and performance characteristics based on whether they are appropriate for their specific uses.

➤ Verifying that utility systems and equipment are built and installed in compliance with the design specifications (e.g., built as designed with proper materials, capacity, and functions, and properly connected and calibrated).

➤ Verifying that utility systems and equipment operate in accordance with the process requirements in all anticipated operating ranges. This should include challenging the equipment or system functions while under load comparable to that expected during normal operation.

It should also include the performance of interventions, stoppage, and start-up as is expected during routine production. Operating ranges should be shown capable of being held as long as would be necessary during routine production. Qualification of utilities and equipment can be covered under individual plans or as part of an overall project plan.

The plan should consider the requirements of use and can incorporate risk management to prioritize certain activities and to identify a level of effort in both the performance and documentation of qualification activities.

Design of facilities and the qualification of utilities and equipment, personnel training and qualification, and verification of material sources (components and container/closures), if not previously accomplished.

Review and approval of the protocol by appropriate departments and the quality unit.. PPQ Protocol Execution and Report Execution of the PPQ protocol should not begin until the protocol has been reviewed and approved by all appropriate departments, including the quality unit. Any departures from the protocol must be made according to established procedure or provisions in the protocol. Such departures must be justified and approved by all appropriate departments and the quality unit before implementation (§ 211.100).

The commercial manufacturing process and routine procedures must be followed during PPQ protocol execution (§§ 211.100(b) and 211.110(a)). The PPQ lots should be manufactured under normal conditions by the personnel routinely expected to perform each step of each unit operation in the process. Normal operating conditions should include the utility systems (e.g., air handling and water purification), material, personnel, environment, and manufacturing procedures. A report documenting and assessing adherence to the written PPQ protocol should be prepared in a timely manner after the completion of the protocol.

This report should:

> Discuss and cross-reference all aspects of the protocol.
> Summarize data collected and analyze the data, as specified by the protocol.
> Evaluate any unexpected observations and additional data not specified in the protocol.
> Summarize and discuss all manufacturing non-conformances such as deviations, aberrant test results, or other information that has bearing on the validity of the process.
> Describe in sufficient detail any corrective actions or changes that should be made to existing procedures and controls.
> State a clear conclusion as to whether the data indicates the process met the conditions established in the protocol and whether the process is considered to be in a state of control. If not, the report should state what should be accomplished before such a conclusion can be reached. This conclusion should be based on a documented justification for the approval of the process, and release of lots produced by it to the market in consideration of the entire compilation of knowledge and information gained from the design stage through the process qualification stage.
> Include all appropriate department and quality unit review and approvals.

Stage 3 — Continued Process Verification

The goal of the third validation stage is continual assurance that the process remains in a state of control (the validated state) during commercial manufacture. A system or systems for detecting unplanned departures from the process as designed is essential to accomplish this goal. Adherence to the CGMP requirements, specifically, the collection and evaluation of information and data about the performance of the process, will allow detection of undesired process variability.

Evaluating the performance of the process identifies problems and determines whether action must be taken to correct, anticipate, and prevent problems so that the process remains in control (§ 211.180(e)). An ongoing program to collect and analyze product and process data that relate to product quality must be established (§ 211.180(e)).

The data collected should include relevant process trends and quality of incoming materials or components, in-process material, and finished products. The data should be statistically trended and reviewed by trained personnel. The information collected should verify that the quality attributes are being appropriately controlled throughout the process. We recommend that a statistician or person with adequate training in statistical process control techniques develop the data collection plan and statistical methods and procedures used in measuring and evaluating process stability and process capability.

Procedures should describe some references that may be useful include the following:

> ➤ ASTM E2281-03 "Standard Practice for Process and Measurement Capability Indices,"

> ➤ ASTM E2500-07 "Standard Guide for Specification, Design, and Verification of Pharmaceutical and Biopharmaceutical Manufacturing Systems and Equipment,"

> ➤ ASTM E2709-09 "Standard Practice for Demonstrating Capability to Comply with a Lot Acceptance Procedure."

Production data should be collected to evaluate process stability and capability. The quality unit should review this information. If properly carried out, these efforts can identify variability in the process and/or signal potential process improvements. Good process design and development should anticipate significant sources of variability and establish appropriate detection, control, and/or mitigation strategies, as well as appropriate alert and action limits. However, a process is likely to encounter sources of variation that were not previously detected or to which the process was not previously exposed. Many tools and techniques, some statistical and others more qualitative, can be used to detect variation, characterize it, and determine the root cause.

Leading manufacturers should use quantitative, statistical methods whenever appropriate and feasible. Scrutiny of intra-batch as well as inter-batch variation is part of a comprehensive continued process verification program under § 211.180(e).

Best practices ensures continued monitoring and sampling of process parameters and quality attributes at the level established during the process qualification stage until sufficient data are available to generate significant variability estimates. These estimates can provide the basis for establishing levels and frequency of routine sampling and monitoring for the particular product and process.

Monitoring can then be adjusted to a statistically appropriate and representative level. Process variability should be periodically assessed and monitoring adjusted accordingly. Variation can also be detected by the timely assessment of defect complaints, out-of specification findings, process deviation reports, process yield variations, batch records, incoming raw material records, and adverse event reports.

Production line operators and quality unit staff should be encouraged to provide feedback on process performance. We recommend that the quality unit meet periodically with production staff to evaluate data, discuss possible trends or undesirable process variation, and coordinate any correction or follow-up actions by production.

Data gathered during this stage might suggest ways to improve and/or optimize the process by altering some aspect of the process or product, such as the operating conditions (ranges and set-points), process controls, component, or in-process material characteristics. A description of the planned change, a well-justified rationale for the change, an implementation plan, and quality unit approval before implementation must be documented (§ 211.100). Depending on how the proposed change might affect product quality, additional process design and process qualification activities could be warranted.

Maintenance of the facility, utilities, and equipment is another important aspect of ensuring that a process remains in control. Once established, qualification status must be maintained through routine monitoring, maintenance, and calibration procedures and schedules (21 CFR part 211, Certain manufacturing changes may call for formal notification to the Agency before implementation, as directed by existing regulations (see, e.g., 21 CFR 314.70 and 601.12).

The equipment and facility qualification data should be assessed periodically to determine whether re-qualification should be performed and the extent of that re-qualification. Maintenance and calibration frequency should be adjusted based on feedback from these activities.

CHAPTER 4

Design Requirements

Requirement definition is the first step in creating a process or piece of equipment that will meet the user's needs and intended use. "Basis of Design" and "User requirement specifications" are documents that are used to formalise specific needs and requirements of a new product, process, facility or utility.

Basis of Design

This is a design document that clearly defines the project and its intended output. It can contain preliminary drawings, key features, technologies, proof-of principle and system descriptions.

User Requirements Specifications (URS)

Requirements should be a specific as possible and should be based upon process knowledge product specifications, regulatory requirements and quality requirements. A User Requirements Document (URS) is the output of the review process and provides a formal approved document for the vendor or manufacturer.

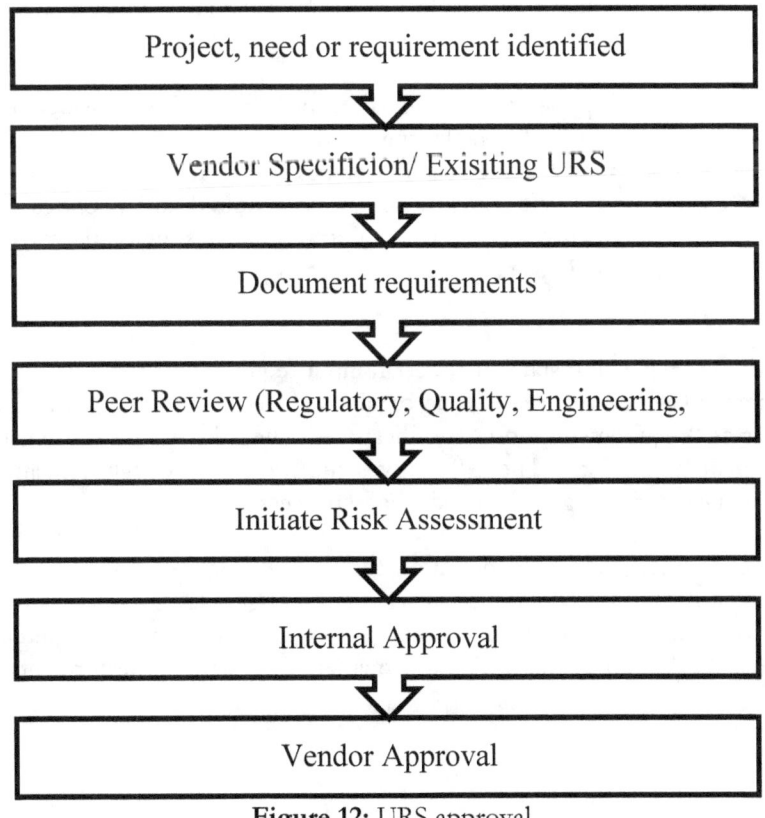

Figure 12: URS approval

When creating a URS it is important to:

> ➤ Specify hardware requirements
> ➤ Describe software requirements
> ➤ Determine the scope of operating parameters
> ➤ Identify safety requirements
> ➤ List calibration requirements
> ➤ Itemise materials of construction (if required)
> ➤ Stipulate the sequence of operation as required

URS documents should be reviewed by appropriately experienced personnel familiar with the product, process and equipment. This ensures critical functions and operations will be identified.

Equipment Classification

Custom: Custom equipment is where the design and requirements are specific to a customer's requirements. Often they are complex as "off-the-shelf" equipment does not meet the customer's needs. With customisation, the potential risk, or potential for manufacturing errors to occur are increased

Off-the-Shelf: Equipment which is available from suppliers that is ready made and does not require customisation.

The second element of Equipment Qualification now must be considered, Equipment-Operational Qualification. This is "Establishing by documented evidence that the equipment operates per specifications and over the required ranges and to required tolerances. Equipment is also tested to ensure alarms and controls operate as required and intended". Some typical checks included in an Equipment-Operational Qualification are testing of alarms, control system testing, utility failures and functional and operational testing.

While Quality System regulation states that design input requirements must be documented, and specified requirements must be verified, the regulation does not further clarify the distinction between the terms "requirement" and "specification." The URS is the starting point for any piece of equipment or machine. The URS drives the project to conclusion and should take account of all the factors relating to the product and process.

A requirement can be any need or expectation for a system or for its software. Requirements reflect the needs of the customer, and may be (1) market-based, (2) contractual, or (3) statutory, as well as a (4) company's internal requirements. There can be many different kinds of requirements (e.g., design, functional, implementation, interface, performance, or physical requirements).

For example, take the scenario where a company wishes to purchase a thermal oven. From a statutory perspective, the company will ensure that the equipment is CE marked and is built in accordance with all EU directives. As this is a fundamental requirement for equipment used within the EU. Similarly, Validation is a statutory requirement for all medical devices and pharmaceuticals. Thus, it will be the responsibility of the customer to validate the equipment and the process. Internal requirements may include the need for a supplier to be certified to a Q.M.S, or that they operate as a limited company.

Software requirements are typically stated in functional terms and are defined, refined, and updated during the development phase. Success in accurately and completely documenting software requirements is a crucial factor in successful validation of the resulting software. A specification* is defined as "a document that states requirements." It may refer to or include Engineering drawings or other relevant documents *21 CFR 820.3(y).

There are different kinds of written specifications:

- User requirements specifications
- System requirements specification
- Software requirements specification
- Software design specification
- Software test specification
- Functional Design Specification

All of these documents establish "specified requirements" and are design outputs for which various forms of Verification or Validation are required. The URS must also define non-software requirements and hardware. Non-functional requirements such as maintainability and usability can also be included. There should be a clear distinction between mandatory regulatory requirements and optional features. The URS should be understood and agreed by both the user and supplier.

User requirements should be:

- Precise
- Clear
- Testable (Verifiable)

Categories

Typically, the following categories are used:

- Equipment layout and siting requirements
- Specification of required utilities
- Facility requirements
- Process and product requirements
- Operational / functional requirements
- Automation/Hardware
- Documentation requirements
- Environmental requirements

Sample User Requirements Specification Template

User Requirement

Overview

This URS is written to document the requirements of *[YOUR EQUIPMENT NAME/PROCESS]*

			Comment	Critical /Non critical
		CONTROL FEATURES		
	3.1	The system cannot begin a cycle until the critical temperatures are reached		
	3.2	The system will automatically start once the program has been selected		
		OPERATIONAL REQUIREMENTS		
	3.3	The machine will not run if any process critical alarms are active		
1.1		The Capacity of the system shall be no less that *[specify requirement]*		
		GENERAL		
1.2		Machine Cycle time shall not be greater than *[specify requirement E.g.10 minutes]*		
1.3		Machine must be operational within *[specify requirement E.g. 10 minutes]*		
1.4		*[Add additional requirements as required]*		
		PROCESS REQUIREMENTS		
2.1		*[Describe the sequence of steps that are required. The example below refers to an injection*		
2.2		Operator selects program		
2.3		Machine reaches operating temperatures		
2.3		Machines starts cycle automatically when critical temperatures are reached		
2.4		Machine executes program		
2.5		Parts are positioned on the outfeed conveyor		

4.1	Modes of Operation are manual and automated		
4.2	On Power failure, no further movement of the equipment will occur without operator		
4.3	E-STOP to be provided within easy reach of the operator position, in accordance with all safety requirements.		
4.4	Visual and Audible alarms must activate if any critical process settings are compromised.		
4.5	Alarms must also appear on the HMI		
4.6	Alarms must remain effective until operator intervention		
4.7	Alarm messages must be in the English language.		
4.8	An alarm log must be automatically stored		
4.9	The vendor must supply a list of alarms indicating failure modes and recovery methods		

	SYSTEM SECURITY			
5.1	HMI must be present on the machine			
5.2	User interfaces must be in the english language			
5.3	Three levels of access required, operator, technician and engineer			
5.4	Each access level will have a different password			
	SAFETY			
6.1	Equipment is "CE" approved			
6.2	The machine meets the European codes of practice (BS or equivalent standards) including; Machinery Directive 2006/42/EC, Low Voltage			
6.3	Guarding over all moving parts			
6.4	Guarding on all live electrical conductors			
6.5	Main Power is fused and has a lock out tag out disconnect			
6.6	Guarding or warning label on pinch points			
6.7	All hazards E.g. hot surfaces are labelled and guarded			
6.8	E-stop push buttons located within reach from normal operator position			
	CALIBRATION			
7.1	System will be calibrated by vendor and calibration certs will be provided			
7.2	Vendor to provide instructions on performing calibrations			
7.3	A Master list of all calibrated instruments will be provided			
7.4	Calibration must be traceable to an external reference standard			

		ENVIRONMENT		
	8.1	Equipment footprint to be no more than approximately 1200mm x 900 mm		
	8.2	Utilities to be supplied to the equipment from the rear.		
	8.3	External surfaces should be compatible with 70% IPA for cleaning purposes		
	8.4	Hazards must be clearly identified		
	8.5	Safety alarms must be visible from operator position		
	8.6	Sound level generated by the equipment must not exceed noise levels of 80 dBA when tested from		
	8.7	If the equipment falls under the category of Electrical and Electronic Equipment as defined in EC Directive 2002/96/EC on Waste Electrical and Electronic Equipment (WEEE) a written declaration stating that the equipment complies with this regulation and EC Directive 2002/95/EC on Restriction of Hazardous Substances in Electrical and Electronic Equipment (RoHS) must be provided by the vendor. A Disposal Manual and WEE label must also be provided for the equipment.		
	8.9	For all agents, oils or processing fluids, an electronic EU format, 16 point MSDS is required for EHS review and approval		
		ERGONOMICS		
	9.1	Controls are easily marked with function		
	9.2	Normal operator position conforms to ergonomic principles e.g. work heights, arm reach, line of sight, seating, foot rests.		

	UTILITIES		
10.1	Machine Requires 415 Volts		
10.2	Frequency 50 Hertz		
10.3	3 Phases		
10.4	Chilled Water?		
10.5	Air Exhaust?		
10.6	Compressed air?		
10.7	Other?		
10.8	The system must not compromised by narrowband interferences (e.g. mobile phones) or broadband interferences as well as not being a source of narrowband or broadband interference that will affect other equipment		
	MAINTENANCE		
11.1	The vendor will recommend a preventive maintenance schedule		
11.2	Vendor to provide a list of recommended held spare parts		
11.3	Maintenance Manuals to be supplied		
11.4	All electrical components and cabling must be identified and labelled correctly on the electrical schematics drawing.		
11.5	All covers to be removable or hinged for easy access.		

	QUALITY			
12.1	Vendor shall operate under a registered Quality System.			
12.2	Vendor shall identify a key contact who will act as Project Manager for the design / development of the equipment.			
	TESTING			
13.1	Vendor is to be responsible for pre delivery testing of the equipment.			
13.2	Factory Acceptance Testing (FAT) to be conducted at the vendor site.			
13.3	Vendor shall take responsibility for any items outstanding from the Factory Acceptance Testing (FAT)			

	DOCUMENTATION REQUIREMENTS		
14.1	The equipment documentation is to be made available for review 4 weeks prior to FAT		
14.2	Electronic copies and 'hard copy' versions of documents to be provided.		
14.3	Control Panel Drawings		
14.4	Equipment Assembly Drawings		
14.5	Electrical Drawings		
14.6	Fixture drawings		
14.7	Spare Parts List and recommended Spare parts list		
14.8	Installation Manuals		
14.9	Operational Manuals		
14.10	Bill of Materials		
14.11	Maintenance Procedure		
14.12	Commissioning Report		
14.13	CE and WEEE Certification		

14.14	Alarm list and guide		
14.15	Declarations of Conformity to EC Machinery Directive 2006/42/EC, Electromagnetic Compatibility Directive EMC 2004/108/EC, Low Voltage Directive 2006/95/EC.		
14.16	Technical drawings & PLC coding copies shall be the property of YOUR COMPANY Operations on completion of the project		
14.17	A list identifying the product contact parts of the machine and the material of construction of such		
	CUSTOMER SERVICE		
15.1	delivery date is 10 weeks after PO placement		
15.2	Vendor shall supply an engineering resource during the development and FAT		
15.3	Vendor to provide training to engineers on installation of the equipment.		
15.4	Post Start-up Support - The machine must remain under warranty for 1 year from date of installation or 2000 operational hours		

APPROVAL		
Author		
System Owner		
Quality		
IT		
EHS		
Facilities		
Validation		

CHAPTER 5

Risk Management

Introduction

Companies operating within the Life sciences typically have a Risk Management strategy applied across their organisations. In simple terms, risk management deals with the identification, assessment, and mitigation risks. Traditionally risk has been described as the "chance or probability of loss". Within the life sciences, potential risk has a greater implication as the welfare of patients can be affected by the use or application of defective products.

The following standards provide a framework for Risk management within a company or organisation.

> ➤ ISO 14971:2012, Risk Management for Medical Devices

> ➤ ISO 13485:2016, Medical Devices Quality Management Systems

> ➤ ICH Harmonised Tripartite Guideline Quality Risk Management, Q9, 2005

Quality risk management is a systematic process for the assessment, control, communication and review of risks to the quality of medicinal products through the product lifecycle. Below a simple model is illustrated. Companies typically customise their Risk management processes to suit their business structure; however, the components identified below are largely included.
The evaluation of the risk to quality should be based on the scientific knowledge and with patient safety as the ultimate goal.

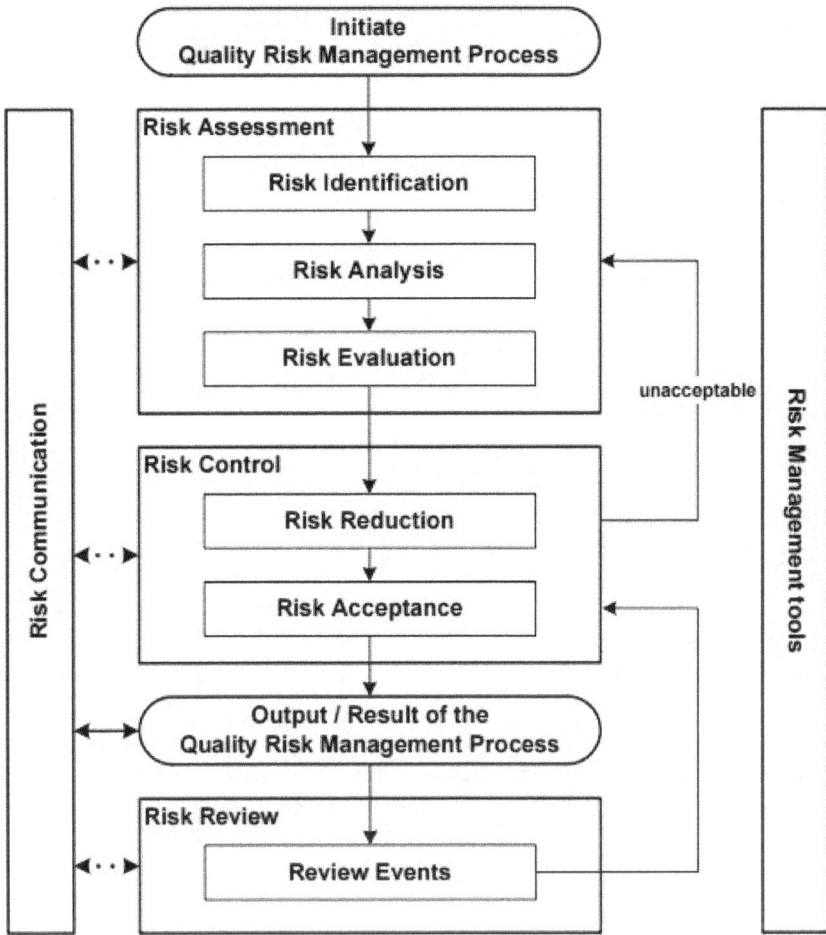

Figure 13: Quality RiskManagement, Adapted from ICH 9

Risk Assessment

Risk assessment consists of the identification of hazards and the analysis and evaluation of risks associated with exposure to those hazards (as defined below). Quality risk assessments begin with a well-defined problem description or risk question. When the risk in question is well defined, an appropriate risk management tool and the types of information needed to address the risk question will be more readily identifiable. As an aid to clearly defining the risk(s) for risk assessment purposes, three fundamental questions are often helpful:

1. What might go wrong?
2. What is the likelihood (probability) it will go wrong?
3. What are the consequences (severity)?

Risk identification

Risk identification is a systematic use of information to identify hazards referring to the risk question or problem description. Information can include historical data, theoretical analysis, informed opinions, and the concerns of stakeholders. Risk identification addresses the "What might go wrong?" question, including identifying the possible consequences. This provides the basis for further steps in the quality risk management process.

Risk analysis

Risk analysis is the estimation of the risk associated with the identified hazards. It is the qualitative or quantitative process of linking the likelihood of occurrence and severity of harms. In some risk management tools, the ability to detect the harm (detectability) also factors in the estimation of risk.

Risk evaluation

Risk evaluation examines the identified and analysed risk against given risk criteria. Risk evaluations consider the strength of evidence for all three of the fundamental questions. In doing an effective risk assessment, the robustness of the data set is important because it determines the quality of the output.

The output of a risk assessment is either a quantitative estimate of risk or a qualitative description of a range of risk. When risk is expressed quantitatively, a numerical probability is used.

Risk can also be expressed using terms such as "high", "medium", or "low", which should be defined in as much detail as possible. Sometimes a "risk score" is used to further define descriptors in risk ranking.

Risk Control

Risk control includes decision making to reduce and/or accept risks. The purpose of risk control is to reduce the risk to an acceptable level. The amount of effort used for risk control should be proportional to the significance of the risk. Decision makers might use different processes, including benefit-cost analysis, for understanding the optimal level of risk control. Risk control might focus on the following questions:

- ➢ Is the risk above an acceptable level?
- ➢ What can be done to reduce or eliminate risks?
- ➢ What is the appropriate balance among benefits, risks and resources?
- ➢ Are new risks introduced as a result of the identified risks being controlled?

Risk reduction

Risk reduction focuses on processes for mitigation or avoidance of quality risk when it exceeds a specified (acceptable) level). Risk reduction might include actions taken to mitigate the severity and probability of harm. Processes that improve the detectability of hazards and quality risks might also be used as part of a risk control strategy.

The implementation of risk reduction measures can introduce new risks into the system or increase the significance of other existing risks.

Hence, it might be appropriate to revisit the risk assessment to identify and evaluate any possible change in risk after implementing a risk reduction process. Risk acceptance is a decision to accept risk.

<u>Risk acceptance</u>

Risk acceptance can be a formal decision to accept the residual risk or it can be a passive decision in which residual risks are not specified. For some types of harms, even the best quality risk management practices might not entirely eliminate risk. In these circumstances, it might be agreed that an appropriate quality risk management strategy has been applied and that quality risk is reduced to a specified (acceptable) level. This (specified) acceptable level will depend on many parameters and should be decided on a case-by-case basis.

Risk Communication

Risk communication is the sharing of information about risk and risk management between the decision makers and others. Parties can communicate at any stage of the risk management process.

The output/result of the quality risk management process should be appropriately communicated and documented. Communications might include those among interested parties; e.g., regulators and industry, industry and the patient, within a company, industry or regulatory authority, etc. The included information might relate to the existence, nature, form, probability, severity, acceptability, control, treatment, detectability or other aspects of risks to quality. Between the industry and regulatory authorities, communication concerning quality risk management decisions might be effected through existing channels as specified in regulations and guidance's.

Risk Assessment Tools

It is best practice to imitate Risk assessments at the commencement of any project. Risk assessments should be developed concurrently with URS' and design documentation in order to capture any potential risks and plan to mitigate against such risks. At the early stages of a project, there may not be adequate information available for a detailed Risk Assessment (e.g. Process FMEA). Therefore, a number of "High Level" risk tools can be used to start the evaluation of risks and get them down on paper.

This section introduces two basic risk assessment tools:

1) High Level Risk Assessment
2) Quality Risk Assessment

High Level Risk Assessment

The following high level risk assessment template can be adapted to suit the particular process, utility or system to be introduced and qualified. It can also be used to establish the GxP status of computerised systems.

Example of High Level Risk Assessment

Equipment /system Details:

System Name		Department	
System No.		Supplier Name	
System Description		System Version	
Document No.		Change Control No.	

Scope:

GxP Risk:

Requirement	Yes	No
The system is part of the environmental controls of GMP areas within the facility?	☐	☐
The system is used during the manufacture of Products?	☐	☐
The system is used for the control of manufactured products?	☐	☐
The system is used for the testing of products?	☐	☐
The computer system is stand alone and is not integrated to other manufacturing processes?	☐	☐
The system is an Off-the-shelf solution or package?	☐	☐
The data generated by the system is used for regulatory submissions	☐	☐

GAMP Category (for Computerised Systems only):

Determine the GAMP Category of the software and system

GAMP Category (1-5)	Category
Comment:	
Is the supplier approved for the GAMP category?	Yes ☐ No ☐

Risk Outcome:

Risk Outcome	Yes	No
The system requires validation based on the potential impact to GxP? (Per section 2)	☐	☐
Based on the GAMP category, the system requires validation (Per section 3)	☐	☐
The system does not require Validation. There is no impact on product quality of GxP?	☐	☐
The overall risk is deemed to be:	Minor ☐ Moderate ☐ Major ☐	
Comment:		

Further Actions:

List actions or mitigations identified during the risk assessment

Action	Completion Date
e.g. Complete Supplier file	

Quality Risk Matrix

A quality risk matrix is an effective way to document high level risks. The structure and format of a typical Quality Risk Matrix is shown below.

Quality Risk Matrix

Quality Risk Matrix

Equipment/System		Part Measurement		Department		Manufacturing
Project No.	#120455	Document No.	QRM-17-005	Change Control No.		1000100

No.	Process Step	Potential Failure	Potential Impact	Risk Likelihood	Severity	Prob. Of Detection	Risk Level H/M/L	Risk Acceptance Y/N	Actions	Risk Control		
										Verification Method	Verification Stage	Verification completed Y/N
1.1	Part Placement	Part not clamped in position	Inaccurate measurement	L	M	H	L	N	1. SOP 2. New fixture 3. Training			

No.	Process Step	Potential Failure	Potential Impact	Risk Likelihood	Severity	Prob. Of Detection	Risk Level H/M/L	Risk Acceptance Y/N	Actions	Risk Control		
										Verification Method	Verificatio n Stage	Verification completed Y/N
1.2	Lighting level setup	Incorrect light level	Reading error	L	L	H	L	N	1. Part specific program 2. Alarm 3. Operat or Training			
1.3	Wrong Program selected	Operator error	Out of spec parts accepted	L	L	H	L	N	1.2nd person verificat ion 2. SOP			

Other common Risk management tools include:

➢ Failure Mode Effects Analysis (FMEA)
➢ Failure Mode, Effects and Criticality Analysis (FMECA)
➢ Fault Tree Analysis (FTA)
➢ Hazard Analysis and Critical Control Points (HACCP)
➢ Hazard Operability Analysis (HAZOP)

FMEA provides for an evaluation of potential failure modes for processes and their likely effect on outcomes and/or product performance. Once failure modes are established, risk reduction can be used to eliminate, contain, reduce or control the potential failures. FMEA relies on product and process understanding.

FMEA methodically breaks down the analysis of complex processes into manageable steps. It is a powerful tool for summarizing the important modes of failure, factors causing these failures and the likely effects of these failures.

FMEA can be used to prioritize risks and monitor the effectiveness of risk control activities. FMEA can be applied to equipment and facilities and might be used to analyze a manufacturing operation and its effect on product or process. It identifies elements/operations within the system that render it vulnerable. The output/ results of FMEA can be used as a basis for design or further analysis or to guide resource deployment.

Failure Mode, Effects and Criticality Analysis (FMECA) FMEA might be extended to incorporate an investigation of the degree of severity of the consequences, their respective probabilities of occurrence, and their detectability, thereby becoming a Failure Mode Effect and Criticality Analysis (FMECA). In order for such an analysis to be performed, the product or process specifications should be established. FMECA can identify places where additional preventive actions might be appropriate to minimize risks. Potential Areas of Use(s) FMECA application in the pharmaceutical industry should mostly be utilized for failures and risks associated with manufacturing processes;

Fault Tree Analysis (FTA) The FTA tool is an approach that assumes failure of the functionality of a product or process. This tool evaluates system (or sub-system) failures one at a time but can combine multiple causes of failure by identifying causal chains. The results are represented pictorially in the form of a tree of fault modes. At each level in the tree, combinations of fault modes are described with logical operators (AND, OR, etc.).

FTA relies on the experts' process understanding to identify causal factors. Potential Areas of Use(s) FTA can be used to establish the pathway to the root cause of the failure. FTA can be used to investigate complaints or deviations in order to fully understand their root cause and to ensure that intended improvements will fully resolve the issue and not lead to other issues (i.e. solve one problem yet cause a different problem). Fault Tree Analysis is an effective tool for evaluating how multiple factors affect a given issue. The output of an FTA includes a visual representation of failure modes. It is useful both for risk assessment and in developing monitoring programs. I.5 Hazard Analysis and Critical Control Points (HACCP) HACCP is a systematic, proactive, and preventive tool for assuring product quality, reliability, and safety. It is a structured approach that applies technical and scientific principles to analyze, evaluate, prevent, and control the risk or adverse consequence(s) of hazard(s) due to the design, development, production, and use of products.

HACCP consists of the following seven steps:

(1) conduct a hazard analysis and identify preventive measures for each step of the process;

(2) determine the critical control points;

(3) establish critical limits;

(4) establish a system to monitor the critical control points;

(5) establish the corrective action to be taken when monitoring indicates that the critical control points are not in a state of control;

(6) establish system to verify that the HACCP system is working effectively;

(7) establish a record-keeping system. Potential Areas of Use(s) HACCP might be used to identify and manage risks associated with physical, chemical and biological hazards (including microbiological contamination).

HACCP is most useful when product and process understanding is sufficiently comprehensive to support identification of critical control points. The output of a HACCP analysis is risk management information that facilitates monitoring of critical points not only in the manufacturing process but also in other life cycle phases.

Hazard Operability Analysis (HAZOP) HAZOP is based on a theory that assumes that risk events are caused by deviations from the design or operating intentions. It is a systematic brainstorming technique for identifying hazards using so-called "guide-words". "Guide-words" (e.g., No, More, Other Than, Part of, etc.) are applied to relevant parameters (e.g., contamination, temperature) to help identify potential deviations from normal use or design intentions.

It often uses a team of people with expertise covering the design of the process or product and its application. Potential Areas of Use(s) HAZOP can be applied to manufacturing processes, including outsourced production and formulation as well as the upstream suppliers, equipment and facilities for drug substances and drug (medicinal) products. It has also been used primarily in the pharmaceutical industry for evaluating process safety hazards. As is the case with HACCP, the output of a HAZOP analysis is a list of critical operations for risk management. This facilitates regular monitoring of critical points in the manufacturing process.

Definitions

Detectability: The ability to discover or determine the existence, presence, or fact of a hazard.

Harm: Damage to health, including the damage that can occur from loss of product quality or availability.

Hazard: The potential source of harm (ISO/IEC Guide 51).

Product Lifecycle: All phases in the life of the product from the initial development through marketing until the product's discontinuation.

CHAPTER 6

Validation Planning

Introduction

Validation planning plays a key role in the qualification and validation of new equipment and processes. They are also used to plan and manage the ongoing validation requirements within companies. So why the need for Validation Plans? Firstly, the requirement for Validation within medical device and pharmaceutical companies is a legal and regulatory one. The Food & Drug Administration (FDA) stipulates Validation as a regulatory requirement of Good Manufacturing Practices (GMP) for both pharmaceuticals (21 CFR 211) and medical devices (21 CFR 820). Validation plans act like a qualification plan that can be used to document strategies, test rationales and key deliverables. They are a regulatory requirement for medicinal products manufactured in the United states and Europe.

Although not stated in 21 CFR Part 820 (Medical devices), validation planning is an important activity that helps to document the validation strategy and deliverables and are common place with medical device manufacturers. In Europe, Eudralex (V4 GMP) is the collection of rules and regulations governing medicinal products in the European Union.

All equipment, processes, facilities and utilities that are GxP impacting need to be qualified. To facilitate the Validation efforts, a Validation Plan (VP) creates a roadmap and structure to meeting the validation requirements. For simple processes or simple equipment repair qualifications a stand-alone Validation Plan may not be required and can be captured within a Protocol or Change Control. The requirements of Validation Plans can be driven by a procedure which may be local to a site or factory, or may be corporate and applicable to multiple sites. Consistency of requirements can also be managed by the use of an approved Validation Plan template.

Generic Benefits

Set aside any regulatory or procedural requirements to create a Validation Plan, there are still valid and beneficial reasons to complete one. A Validation Plan acts as a top level document that can pull together the many references, protocols, reports and rationales that make up a validation project.

It is also a powerful asset to introduce new staff and team members to a project or process who need to get up to speed quickly and comprehensively. Validation Plans are often the first documents an auditor will request to see in relation to a new process or new product introduction. Validation Plans force the various stakeholders to sit down and agree upon the strategy and any technical rationales required to deliver a successful Validation.

Types of Validation Plans:

- ➢ Site Master Validation Plan
- ➢ Master Validation Plan
- ➢ Individual Validation Plan

Site Master Validation Plan

A Site MVP details the Products, Processes and associated Validation Protocols and reports for a Manufacturing site/factory. It is the over-archiving validation plan. Typical Site MVP includes: Description of products & processes, test methods (analytical, physical), specifications, an up-to-date list of utility qualifications, equipment qualifications, process validations.

Master Validation Plan

A MVP encompasses all aspects of a validation strategy where multiple processes, equipment/machines require validation. MVP s are common for New Product Introductions. Although not stated in 21 CFR Part 820 MVPs are an important requirement to document the validation status.

Individual Validation Plan

Individual Validation Plans document the validation strategy or general approach to qualification for a particular system, piece of equipment or Process. Individual Validation Plans are also used for Projects which fall outside of the day to day validation requirements.

For simple equipment the Validation Plan may be documented in the Qualification protocol.

Matrix or Family Approaches

A family or matrix approach to validation can be used where similar products are produced using the same equipment and processes. A particular product size or product configuration may be selected to represent the "worst-case" product. Therefore, by qualifying the worst case, all of the other products within the family are considered validated. Matrix or Family approaches must be clearly documented and technical rationale provided in advance of any qualification activities. This can be addressed in a Validation Plan or within a Protocol. Alternatively, a technical report of P/dev report can be created and referenced.

Stages of Commissioning, Qualification and Validation

Many companies create a bespoke in-house process of the Commissioning, Qualification and Validation Process. It can be represented in 3 steps (C/Q/V) or further divided or sub-divided to suite the organisational structure or corporate requirements. Below, a 4 step process covering the validation lifecycle is described.

- ➢ Stage 1 (Design)
- ➢ Stage 2 (Planning, Commissioning & Qualification)
- ➢ Stage 3 (Validation)
- ➢ Stage 4 (Continued Verification /Decommissioning)

Stage 1 (Design)

The introduction new product, process or process improvement program within the life sciences (medical device, pharmaceutical and biopharmaceutical etc.) must be done in a well-managed and defined process. A risk based and process approach is the gold standard and gives a degree of assurance that a project reaches its intended goal.

The relevant stake-holders and expertise must be a part of the design stage and should give adequate consideration to regulatory requirements, industry standards, GxP, GEP and company in-house requirements.

Requirements and the intended use of the equipment/systems are typically documented in a User Requirement Specification (URS). The purpose of a URS is to document the general requirements and critical requirements of new equipment and systems and its intended use.

Largely speaking, a URS is an engineering document and as such it normally created by a Process Engineer or Process owner. Principally they are responsible for ensuring the specification and design meets fulfils the operational and functional requirements. Processes that are critical to product quality and patient safety must be assured to be controlled with automation and other in-process checks. Other functions such as validation, H&S, quality, regulatory and automation should also be involved with new equipment and systems.

As the project develops, design reviews are typically completed during the course of the design development. The manufacturer or vendor needs to interpret the requirements and contents of the URS. Design reviews are documented formal reviews that occur prior to Design Qualification (DQ). For example, a design review may be completed prior to the placement of an order, or in response to a vendor quotation or scope document.

Stage 2- Planning of Commissioning/Qualification

Commissioning or Qualification plans for GMP equipment, facilities, utilities and systems should be generated. A Qualification Plan (QP) describes all the qualification measures and at which stage of the qualification the verification will be completed.

A Qualification Plan typically contains detailed descriptions of the necessary test measures and a description of the interdependencies of the individual tests. References to other test documents such as FAT or SAT and a description of the deviation management may also be integrated in the Qualification Plan.

In some instances, there may not be a need or a requirement for a Qualification Plan. A Validation Plan can also serve to detail the Commissioning and Qualification strategy if permitted by the internal procedures within a company. A simple table or matrix can be used to map out the requirements and qualification activities.

Test	FAT	COMISSIONING	SAT	IQ/OQ
Drawings review	X			
Documentation	X			X
Calibration		X		X
Walk down	X		X	
Alarm testing	X			X
Functional testing	X		X	X
Utilties		X	X	

Figure 14: Simple Qualification Plan.

GxP Assessments

Equipment, processes, utilities and facilities need to be assessed in order to identify if they are GxP impacting or applicable. Formal GxP assessments should be completed to determine if a system has a direct impact, indirect impact or no impact on the quality of product. A system that does not impact the quality of safety of a product can be classified as a non-GxP system.

Commissioning and Qualification

Commissioning and qualification activities include the execution of FATs and SATs of equipment, systems or utilities. While they are typically classified as engineering documents, they should be completed using pre-approved protocols. As most equipment and systems are GxP impacting, Good Documentation Practices (GDP) and Good Engineering Practices (GEP) should be adhered to. Thought should also be given to the approval requirements. A small number of approval signatures on a protocol may allow a quick review and swift approval, however, it may lead to issues later if the executed protocol is to be leveraged. Some companies may not require a Quality approval on engineering protocols, however, including quality review and approval at the commissioning stage of a project helps to ensure tests are aligned with site norms and expectations.

In strict terms, commissioning activities can focus on both GMP and non-GMP requirements of equipment and systems etc., while qualification must include regulatory requirements and documents that critical aspects and regulatory requirements meets their acceptance criteria. The URS should provide the basis of testing requirements along with critical process parameters and other requirements identified in Risk assessments. Leveraging of FAT/SAT activities in order to reduce the qualification effort is a common approach and if it is properly risk based the testing should be sufficient.

Installation Qualification (IQ) ensures that the equipment or system is installed in accordance with manufacturers recommendations:

> Installation of equipment, piping, utilities, and instrumentation checked to current engineering drawings and specifications and ensuring drawings are verified to 'as-built' state.
> Logbooks (or electronic equivalent controls) have been established.
> Supplier and site operating and working instructions and maintenance requirements have been collected and are available and that the equipment has been entered into the preventive maintenance program.
> Verification that instruments have been classified and calibrated
> Verification of materials of construction.

Operational testing (OQ)

> Functionality of the facilities, equipment, utilities, systems, or processes has been demonstrated to meet predefined acceptance criteria.
> Alarm testing has been demonstrated to meet predefined acceptance criteria.
> Control system functionality has been demonstrated to meet predefined acceptance criteria.
> Equipment control within the process ranges for CPPs has been demonstrated to meet predefined acceptance criteria.
> Operating procedures, cleaning or sanitization procedures, and preventive maintenance procedures are accurate and ready for approval prior to PQ.

Testing at IQ and OQ level may be leveraged from Commissioning (including FAT, SAT) if fully documented and planned for at the commissioning stage.

Stage 3 -Validation

The aim of PQ activities is to establish confidence testing that the process / equipment meets all release requirements for functionality and the manufacture of product is safe, consistent and repeatable. The PQ approach and associated activities are typically outlined in the validation plan for the system / equipment / product being qualified.

It should be noted that PQ execution can be combined with IQ (Installation Qualification) and OQ (Operational Qualification) activities depending on the overall approach, internal procedures and the complexity of the equipment of process. Complex systems are typically done step by step. Where a simple process exists validation may incorporate equipment and process validation into one document.

Stage 4 (Continued evaluation/Discontinuation)

A Continued Process Validation Program is put in place to further verify the validated state through the life of the equipment or process.

Changes to the Validated State

Revalidation may be necessary under such conditions as:

> ➤ change(s) in the actual process that may affect quality or its validation status
> ➤ change(s) in the product design which affects the process
> ➤ transfer of processes from one facility to another
> ➤ change of the application of the process

The need for revalidation should be evaluated and documented. Evaluation should consider historical results from quality indicators, product changes, process changes, changes in external requirements (regulations or standards) and other such circumstances, as applicable. Revalidation may not be as extensive as the initial validation if the situation does not require that all aspects of the original validation be repeated.

Changes impacting Operational Qualification- OQ

For changes made to a qualified process, evaluate whether worst case conditions exist.
If no worst-case conditions exist for the affected outputs, then an OQ of the change is not required. Rationale for not conducting an OQ shall be documented in the PQ report. While narrowing or tightening process parameters within a qualified range does not typically require qualification. Take a scenario that seeks to Widen Process Parameter Outside of the Qualified Range. Suggested minimum validation activity would include:

> ➤ Both Operational and Performance Qualification runs to quality the new range.
> ➤ All product outputs impacted by the change should be tested and challenged as would happen in an initial validation.

Changes impacting Performance Qualification-PQ

If changes are proposed to a qualified process, the impact of the changes should be assessed and documented. Changes to critical process parameters may require a full re-qualification depending on the level of change that is proposed. Supporting studies may be required to support the proposed change and such as Engineering studies or testing.

Change Control

Validation programs are subject to change control. Each company or organisation should have a procedure detailing the change management process. Below is a suggested overview of a typical change control process. Any system, facility, document or process that has the potential to impact product quality and validated state is generally subject to following a change control process. Another term used in industry is Enterprise Change control or Engineering Change control. Essentially these terms are the same. The intent is to control and manage change consistently.

A change control can take a form of a document which drives the agenda and the specific requirement. Change control are also created with Enterprise software such as Kintana, Documentum and SAP to name but a few. While each company will have varying processes, some basics are common. These include the 3 stages of change Control-Pre-implementation, implementation and Post implementation (if required). Below, 2 case studies are detailed where there is a change in manufacturing which requires a formal change control process to be applied.

CHAPTER 7

Clean Utilities

Introduction

The term "Clean Utilities" in the life science industry refers to utilities that have to fulfil regulatory requirements. The most common utility is water, which can be supplied in different pharmaceutical grades of purity. Purified water (PW or PUW), Highly Purified Water (HPW) and Water for Injection (WFI) are the most common. Water quality specifications can be found in the pharmacopaeias, e.g. the US Pharmacopeia. Other clean utilities can also include clean compressed air, clean gasses (e.g. nitrogen, argon and oxygen), and clean steam.

Key Definitions

Alert limit: a value reached when the normal operating range of a critical parameter has been exceeded, indicating that corrective measures may need to be taken to prevent the action limit being reached.

At-rest: a condition where the installation is complete with equipment installed and operating in a manner agreed upon by the customer and supplier, but with no personnel present.

Cleanroom: an area (or room or zone) with defined environmental control of particulate and microbial contamination, constructed and used in such a way as to reduce the introduction, generation and retention of contaminants within the area.

Containment: a process or device to contain product, dust or contaminants in one zone, preventing it from escaping to another zone.

Contamination: the undesired introduction of impurities of a chemical or microbial nature, or of foreign matter, into or onto a starting material or intermediate, during production, sampling, packaging or repackaging, storage or transport.

Point extraction: air extraction to remove dust with the extraction point located as close as possible to the source of the dust.

Pressure cascade: a process whereby air flows from one area, which is maintained at a higher pressure, to another area at a lower pressure.

Relative humidity: the ratio of the actual water vapour pressure of the air to the saturated water vapour pressure of the air at the same temperature expressed as a percentage. More simply put, it is the ratio of the mass of moisture in the air, relative to the mass at 100% moisture saturation, at a given temperature.

Turbulent flow: turbulent flow, or non-unidirectional airflow, is air distribution that is introduced into the controlled space and then mixes with room air by means of induction.

Identifying Critical Utilities

The process of identifying critical utilities can be done with the application of direct impact, indirect impact and no impact definitions (see previous section "*risk and impact assessment*"). Risk assessments, CQAs and CPPs should also help identify critical utilities. When critical utilities are required as part of manufacturing and processing, the following points should be examined during the requirements and design stage:

- Materials of construction
- Internal surface finishes
- System sizing
- Flow rates, dead legs, drainage etc.

The process of identifying critical utilities can be done with the application of direct impact, indirect impact and no impact definitions (see previous chapter). Risk assessments, CQAs and CPPs should also help identify critical utilities. When critical utilities are required as part of manufacturing and processing, the following points should be examined during the requirements and design stage:
- ➤ Materials of construction
- ➤ Internal surface finishes
- ➤ System sizing
- ➤ Flow rates, dead legs, drainage etc.

Compressed Air

Compressed air is used for valve actuation, instrument air and process air to name but a few applications. Only the point-of-use filtration and the gas quality instrumentation should be classified as level 1. When flow or pressure is a CPP, the measurement/monitoring should be performed by the system into which the gas is flowing. Additionally, the CQAs and CPPs should be routinely monitored through the calibrated monitoring system. For compressed air, the potential CPPs are listed below. For the physical system being evaluated, the use and the application of the compressed air will determine which (if not all) CPPs are needed to ensure the system produces product of the desired quality.
- ➤ Hydrocarbons
- ➤ Moisture
- ➤ Particulates
- ➤ Temperature

It is important that each point of use has appropriate sterile filters in place. If the filter is not placed directly at the point of use, control and counter measures should be implemented to address any risk of contamination downstream of the filter. Compressed air for biopharmaceutical use must be generated using oil free compressors with appropriate temperature controls in place.

Water Systems

Water supply and the associated Water Systems in biotechnology and pharmaceuticals is a vital component of the manufacturing process. It is used to clean equipment and vessels, to cool or heat processing pipes and systems, and in many circumstances certain grades of water is a component of the finished product (e.g. Water-for-injection). Various grades of water service a particular purpose. Some common types include:

> Potable water
> Soft water
> Purified water
> Water-for Injection

Water used in process and in cleaning should be pure and free from microbial and chemical impurities. As the water gets easily contaminated by environmental conditions, diligence in the design is essential. Typically water systems are supplied on a continuous loop with recirculation.

CPPs typical for a water system include:
> Pressure
> pH
> Conductivity
> Level
> TOC
> Flow
> Temperature
> Resistivity

Water for Injection

The use of WFI is two-fold. Firstly it can be used for critical processing steps such as washing and rinsing .It can also be used in Injectable products. WFI is a key raw material for sterile intravenous and intradermal products. WFI is produced by Multi Column Distillation Plant (MCDP), and must meet the microbial requirements of regulated bodies.

Clean-in-place (CIP) / Sterilise-in-place (SIP) system

The cleaning of equipment, vessels and process piping is a critical activity. Any residue from a previous production batch needs to be removed in order to avoid cross contamination. CIP and SIP skids often utilised to allow efficient switchover between batches and/or products.

Clean steam

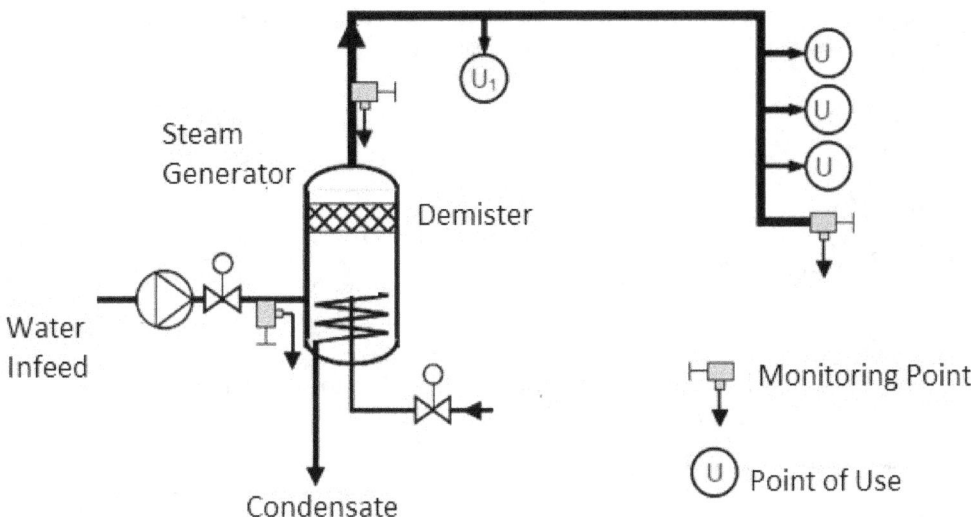

Figure 15: Simple Clean Steam Generation Piping and Instrumentation

Pure Steam is used in Pharma and Biotech for sterile application, for Autoclave sterilization etc. Distribution piping of Clean Steam is a critical aspect. Improper sizing of pipes may lead impact the production process and loss of time during sterilization.

Clean steam, also referred to as "pure steam", and gases used in manufacturing operations must be of a quality suitable for their intended purpose. The intended use of Clean steam and gases must be understood in order to determine any risks to the patient or product. For example gases end up being part of the product must fulfil the regulatory requirements. Preventative Maintenance and on-going monitoring must be implemented for Clean Steam systems.

➢ Routine inspection and maintenance.
➢ Frequency of filter change
➢ Frequency of the sterilization for the gas distribution system, if applicable
➢ Frequency for integrity testing of the sterile filter

Water systems for Purified Water, De-ionised water and Water for Injection (WFI) must provide a consistent and reproducible output. Where there is moisture, there is always a risk of microbial contamination. Therefore, the design of water systems should mitigate against such risks. Good Engineering practices such as using circulation loops, no dead legs and polished surface finishes all work to provide an effective and safe system. The design should also take into account ease of sampling at the point of use. The removal of endotoxins is a requirement for WFI.

On-going sampling monitoring the quality of water is particularly important when water systems are concerned. Procedures should be in place to ensure effective monitoring and testing is maintained. Action limits and acceptance criteria should be clearly documented in approved SOPs or equivalent. Failure to meet limits or acceptance criteria should initiate an investigation. The potential CPPs are listed below for Clean Steam Systems:

- Conductivity
- Flow
- Level
- Pressure
- Resistivity
- Temperature

Design Considerations

The purpose of a User Requirement Specification (URS) is to define the requirements for the operation and control of the clean steam system.

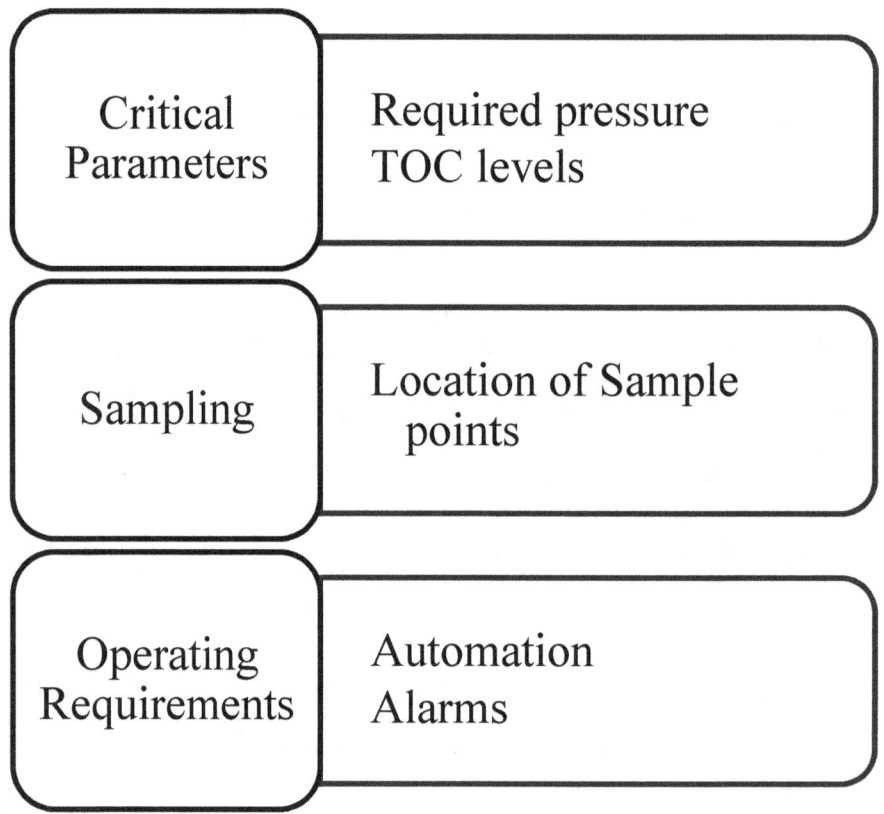

HVAC

Heating, ventilation and air-conditioning (HVAC) plays an important role in ensuring the manufacture of quality products. Furthermore, HVAC systems also provide comfortable conditions for operators based in the manufacturing environment. HVAC system design influences the layout of airlock positions and doorways. In turn, airlocks, entrances and exits have an effect on room pressure differential cascades and cross-contamination control. The prevention of contamination and cross-contamination is an essential design consideration of the HVAC system. In view of these critical aspects, the design of the HVAC system should be considered at the concept design stage of a manufacturing plant. Temperature, relative humidity (RH) and ventilation should not adversely affect the quality of products during their manufacture and storage, or the proper functioning of equipment. CPPs for HVAC systems include:

> ➤ Temperature
> ➤ Humidity
> ➤ Particle count – viable and non-viable
> ➤ HEPA filter certification/leak test/air flow rates
> ➤ Room differential pressures

The Displacement Concept (low pressure differential, high airflow)

This concept is commonly found in production processes where large amounts of dust are generated. Under this concept the air should be supplied to the corridor, flow through the doorway, and be extracted from the back of the cubicle. Normally the cubicle door should be closed and the air should enter the cubicle through a door grille, although the concept can be applied to an opening without a door. The velocity should be high enough to prevent turbulence within the doorway resulting in dust escaping. This displacement airflow should be calculated as the product of the door area and the velocity, which generally results in relatively large air quantities.

Note: This method of containment is not the preferred method, as the measurement and monitoring of airflow velocities in doorways is difficult.

Pressure Differential Concept (high pressure differential, low airflow)

The pressure differential concept may normally be used in zones where little or no dust is being generated. It may be used alone or in combination with other containment control such as a double door airlock. The high pressure differential between the clean and less clean zones should be generated by leakage through the gaps of the closed doors to the cubicle. The pressure differential should be of sufficient magnitude to ensure containment and prevention of flow reversal, but should not be so high as to create turbulence problems.

In considering room pressure differentials, transient variations, such as machine extract systems, should be taken into consideration. A pressure differential of 15 Pa is often used for achieving containment between two adjacent zones, but pressure differentials of between 5 Pa and 20 Pa may be acceptable. Where the design pressure differential is too low and tolerances are at opposite extremities, a flow reversal can take place. For example, where a control tolerance of $\pm$ 3 Pa is specified, the implications of rooms being operated at the upper and lower tolerances should be evaluated. It is important to select pressures and tolerances such that a flow reversal is unlikely to occur. The pressure differential between adjacent rooms could be considered a critical parameter, depending on the outcome of risk analysis.

The limits for the pressure differential between adjacent areas should be such that there is no risk of overlap in the acceptable operating range, e.g. 5 Pa to 15 Pa in one room and 15 Pa to 30 Pa in an adjacent room, resulting in the failure of the pressure cascade, where the first room is at the maximum pressure limit and the second room is at its minimum pressure limit. Low pressure differentials may be acceptable when airlocks (pressure sinks or pressure bubbles) are used to segregate areas.

The pressure control and monitoring devices used should be calibrated and qualified. Compliance with specifications should be regularly verified and the results recorded. Pressure control devices should be linked to an alarm system set according to the levels determined by a risk analysis. Manual control systems, where used, should be set up during commissioning, with set points marked, and should not change unless other system conditions change. Airlocks can be important components in setting up and maintaining pressure cascade systems and also to limit cross-contamination. Airlocks with different pressure cascade regimes include the cascade airlock, sink airlock and bubble airlock.

CHAPTER 8

Equipment Validation

Introduction

Validation is a legal and regulatory requirement for the manufacture of medicinal products. The area of Validation can be sub-divided into two elements. Equipment Qualification (EQ) and Process Validation. Equipment qualification ensure that equipment operates as intended and is installed in accordance with the manufacturers recommendation. Process Validation involves the provision of documented evidence to confirm a particular process performs consistency and meets pre-determined specifications.

All equipment that can impact the quality of product is subject to Validation, hence equipment and systems used in aseptic manufacturing must undergo equipment and process validation. Installation Qualification (IQ) protocols should cover verification that all utilities are installed correctly to the manufacturers recommendations. All sitting and mechanical connections should also be confirmed as adequate. Other key tests and verifications includes:

➢ Documentation of Materials of Construction (MOC)
➢ Calibration of equipment based instrumentation
➢ Spare parts listing
➢ Preventative maintenance schedule creation
➢ Electrical installation verification
➢ Health and Safety assessment
➢ Ergonomic Assessment
➢ Documenting Software and Hardware
➢ Backup of software
➢ Backup of Recipes (Sterilization, Bio-decontamination etc.)

The system User Requirements Specification (URS) should provide the basis of testing and must be fulfilled during the course of Validation.

The ultimate goal of Equipment Qualification is to ensure that equipment is fit for its intended use. Therefore, equipment is validated to confirm it functions as intended and meets all requirements to manufacture product safely and consistently. FDA requires that "Each manufacturer shall ensure that all equipment used in the manufacturing process meets specified requirements and is appropriately designed, constructed, placed and installed to facilitate maintenance, adjustment, cleaning and use". In other words all manufacturing equipment, support facilities, measuring and test equipment shall be "qualified". (FDA 21 CFR 820.70 (g))

Equipment Qualification Protocols are developed to document this testing and hence provide evidence on the functionality and consistency of the equipment. There are two distinct parts within the scope of Equipment Qualification, Installation Qualification and Operational Qualification. Often these subparts are abbreviated to IQ and OQ. Other combinations such as IOQE and IQ/OQ can be encountered within industry. This is often defined in a company's procedure or SOP relating to Equipment Validation.

A User Requirements Specification (URS) is often used to document the "specified requirements" of a particular piece of equipment. A URS can then be used as an input document when equipment qualification is required. While a URS document can be extensive covering areas such as equipment functionality, utility requirements, safety features, software specs etc. not all requirements documented in a URS will need to be verified or validated. Critical requirements should be identified early and should always be verified.

In short Equipment Qualification is confirmation via documented evidence that the particular requirements for a specific intended use can be consistently fulfilled under anticipated conditions.

Often referred to as the three "C" s of Validation – confirmation – consistency and conditions (anticipated). These are key themes that Validation must address. Confirmation is addressed by the process of completing a formal validation. When it's done, it is documented and available for review to auditors. To assess consistency, there must be a number of batches or "runs". Typically, there is minor batch-to-batch differences or variations between batches. These differences can be as a result of setup or raw material differences. Process Validation must ensure that despite minor changes, there is consistency between batches, with product meeting specifications. Controlled or anticipated conditions are the machine or process settings that are known, documented and controlled during the manufacture of products.

Materials of Construction (MOC)

The materials of construction and evidence of the same (certificates) forms part of Installation also. Materials must be fit for the intended purpose and compatible with products and manufacturing agents that come into contact with them.

For example, fermenters are made of materials that are suited to the use of steam sterilisation techniques and regular cleaning. Such materials can be classed as both non-reactive and non-absorptive surfaces. Most aseptic processing equipment that incorporates product contact surfaces are made of high grade stainless steel. Cheaper classifications of stainless steel can be used for jacketing and other non-product contact areas.

All interior product contact surfaces should be polished to a "mirror" finish. Welds also need to be finished in a similar manner. Electro polishing provides a better quality surface finish than mechanical polishing.

As with any chemical reaction, factors such as temperature, pH and oxygen concentration can impact the performance and yield. To ensure the optimum conditions are maintained, it is important to monitor and control such parameters and factors. By far the most common these days is automatic control of systems and equipment with automatic feedback and adjustment.

Operational Qualification (OQ) is the second component of Equipment Qualification. This is "Establishing by documented evidence that the equipment operates per specifications and over the required ranges and to required tolerances". Equipment is also tested to ensure alarms and controls operate as required and intended. Some typical checks included in an equipment-operational qualification are testing of alarms, control system testing, utility failures and functional and operational testing.

Suggested IQ/OQ Verifications / Tests

Standing Operating Procedures (SOPs)

SOPs are designed to provide formal documented instruction on how to execute tasks or operate equipment or machinery. While each company will require different headings, a work instruction or SOPs should cover set-up, system operation, cleaning and shutdown to name but a few.

Test Instrumentation Calibration

External test devices such as temperature probes, volt meters, lux meters and particle counters may be required to take measurements during an equipment qualification. Test instrumentation should have a suitable range, resolution and accuracy. Certificates of calibration should also be available, with calibration conforming to a recognized external standard. Information such as the serial number, model number and manufacturer should be recorded for reference. And traceability.

Equipment Based Instrumentation Calibration

All equipment based instruments must be calibrated as part of equipment validation or in advance of it. Instruments should have unique calibration ID.

Electrical Checks

Appropriate connections and earthing checks. Review of electrical drawings to ensure the physical status is as per drawings and specifications. Cables and electrical hazards should be also appropriately labelled.

Mechanical Checks

Ensure the systems are fixed, fastened and integrated mechanically. Safety guards and barriers should also be in place where required.

Pneumatic Checks

Verify the proper supply and integration of compressed air. Supply should be leak free, regulated with filters and watertraps fitted as required.

Documentation

Verification that design, operation and maintenance documentation has been received from the manufacturer and are stored appropriately.

Ergonomics

Controls and HMIs should be positioned to facilitate ease of use and should be identified clearly.

Health and Safety

Hazards are identified and guarded, pin points are identified. No trip hazards are evident and Emergency stops function.

Software

Equipment Installation software checks should record the names and version numbers of all software. HMI Software, PLC software, application software. Provision should be made for disaster recovery and backup.

Hardware

Computer hardware should be recorded to include the model, manufacturer, serial number and specification details.

Environmental

Any features detailed in the URS relating to environmental requirements need to be verified during IQ/OQ. For example, automatic shutdown after periods of inactivity. Heating and cooling systems should also be appropriately insulated.

Alarms

Automated processes such as sterilization tunnels, Autoclaves, Filling machines and Isolators typically have many alarms and controls. Alarms can be categorized as critical or non-critical to the process or product. Depending on the vendor or manufacturer alarms can also be grouped according to the type of alarm (EHS, process, mechanical, pneumatic and so on) Alarms should be tested to ensure the right action by the machine is taken, the process comes to a safe stop, and that the alarm can be acknowledged and the alarm condition cleared.

Utility Failure

Also referred to as provoke testing, utility failure of compressed air, fume extraction, electrical supply and so is to ensure in the event of failure during commercial manufacturing, the equipment comes to a safe stop and can be brought back into use upon recovery of the utility.

Fixtures

Materials of construction must be suitable for the intended use. In aseptic processing, high grades of stainless steel (316L) are the preferred material of use.

Functional Tests

The individual functions of equipment must be verified during commissioning and qualification.

Suggested Pre-Requisites to Equipment Qualification

Prior to formal Equipment Installation and Operational Qualification (IQOQ), there are a number of engineering activities that can be completed. Although work is required upfront in order to complete these activities such as preparing engineering test protocols, they will benefit the qualification stage. The completion of some or all of the above activities will help identify issues prior to formal equipment qualification. Essentially, an FAT Protocol is like an early draft of an IQOQ-Equipment Protocol.

Factory Acceptance Testing (FAT)

FAT or Factory Acceptance Test – is an Engineering activity, the purpose of the FAT is to verify the equipment or system meets the requirements of the URS. From the Validation Engineers perspective, it can be a learning activity and an opportunity to gather data, documentation and supporting design documents that will prove valuable during the equipment and Process Validation of the equipment. SAT is an Engineering activity that is completed at the site of the vendor or equipment manufacturer, post FAT.

Site Acceptance Testing (SAT)

Site Acceptance testing is also an engineering activity conducted when equipment arrives onsite. Depending on the company, SAT can be completed by the vendor or by the purchasing company. It consists of a series of installation and operational checks to ensure the equipment has not suffered any damage or deterioration between the disassembly, crating, shipping and delivery of the equipment.

Equipment Qualification (EQ) Protocols

Protocol Preparation

Thorough and careful preparation is critical in order to successfully complete a qualification without deviations. In the preparation of the protocol, the URS is a key document. Often quotations, design documents, vendor drawings and owner manuals can contribute to the test and verifications to be completed during EQ.

Protocol Approval

Approval is always required prior to executing an equipment qualification protocol. Approvers should be aware that they are signing for the accuracy and content of the whole document. It is strongly advised that prior to final approval and execution of a protocol, a dry-run or trial is completed to ensure the test methods and acceptance criteria are accurate.

Post Execution Review

Upon execution of a protocol, timely review is advised in order to catch any errors or omissions. The person reviewing the protocol should not be the same person who performed the test. This review is best completed by a Quality Engineer; however, each organisation should identify personnel responsible for EQ reviews.

Some points to remember when completing protocols:
- Ensure the protocol is fully approved prior to execution
- Ensure personnel are trained on the protocol (if required) and trained to the specific work instructions
- Ensure all team members, contractors etc. sign the signature log
- Observe safety precautions and wear PPE as required
- Ensure other employees that may be impacted by qualifications are aware that a qualification is in progress.
- Ensure that any test product required in support of the qualification is identified segregated and stored to internal standards
- Check that accurate work instructions are available (some companies may allow redlined copies to be used)
- Complete all tests in the protocol
- Always use indelible ink
- Carefully check each result against any acceptance criteria
- Ensure all test equipment is validated within calibration prior to use
- Handwritten comments should be signed and dated per GDP
- Deviations should be written up at the time of observation
- Data records and attachments should be identified with the protocol number, signed, paginated and dated
- When data is transcribed it should be verified by a second person. The source of the data should also be recorded
- All product manufactured should have relevant batch documentation as per normal production conditions

Equipment Qualification Reports

On completion, Equipment Installation and Operational Protocol reports are required. The format of any report largely depends on company specific procedures. If the protocol is an executable document (results are hand written in) then the executed version can be deemed the report. A summary report may be required but this depends on the requirements within your company or organisation.

The typical requirements of a completed EQ validation include:

- Equipment Qualification Protocol
- Equipment Qualification Protocol (executed)
- Raw Data
- Attachments (Examples of attachments include –material certs, calibration certs, CE Cert and MSDS'.

Software Validation

Where there is the potential to affect product conformance to requirements or where software or IT systems provide support to aspects of Quality Management, validation is required.

Most companies categorise software validations to account for the different applications of software and IT systems. For example, Enterprise systems, such as the drawing package SolidWorks would be validated in a different manner to Manufacturing Systems that contain software (a.k.a. embedded software).

"Embedded" software is where the software is integrated into the manufacturing equipment. Embedded software is typically validated during the Equipment Qualification stage, Process Validation stage or Test Method Validation. Enterprise software falls outside of Equipment or Process Validation but does require validation if it impacts product quality or is used to make quality decisions. Standalone systems such as ERP (Enterprise Resource Planning) systems also require validation.

Software Validation & GAMP

Good Automated Manufacturing Practice (GAMP) is a set of guidelines for manufacturers and users of automated systems in regulated industries. Specifically, the Medical device, pharmaceutical and biopharmaceutical industries. The application of GAMP and Validation of Automated Systems in manufacturing helps ensure that regulated medical devices and medicinal products have the required quality and are manufactured according to Good practices, meet regulatory and legal requirements and ensure patient safety. GAMP ensures quality is in-built into each stage of the manufacturing process. Therefore, GAMP has a place in all aspects of automation and production, including the handling of raw materials, control of facilities and equipment etc.

Key Terms

Automated System: Term used to cover a broad range of systems, including automated manufacturing equipment, control systems, automated laboratory systems manufacturing execution systems and computers running laboratory or manufacturing database systems. The automated system consists of the hardware, software and network components, together with the controlled functions and associated documentation. Automated systems are sometimes referred to as computerised systems; in this Guide the two terms are synonymous.

Commercial off-the-shelf (COTS): Configurable Programs- Stock programs that can be configured to specific user applications by "filling in the blanks", without (COTS) altering the basic program.

Computer System Validation: a process that confirms by examination and provision of objective evidence that the computer system conforms to user needs and intended uses. System validation is a process for achieving and maintaining compliance with GxP regulations and fitness for intended use by adoption of life cycle activities, deliverables, and controls.

GAMP 5: is a set of guidelines that offers a Risk-Based approach to ensuring the compliance of GxP impacting computerised systems.

V- Model: is a development process which sets out a roadmap of stages and deliverables during a project.

21 CFR Part 820: FDA requirements pertaining to Medical Devices.

User Requirement Specification, URS: The URS is a critical document that defines the requirements of the computerised system and agreement to the requirements.

Software Requirement Specification, SRS: an SRS can be written to interpret the requirements of a URS and how they relate to the requirement or how the requirement is met in practical terms regarding software.

Functional Design Specification, FDS: a functional design specification is a document that specifies how particular requirements are met – this can be a combination of how the equipment/process operates mechanically/automatically etc. An FDS is typically written to response to a URS.

Computer System Validation Life Cycle

The Computer System Validation Life Cycle refers to all activities from initial concept to retirement of a computer system. The life-cycle of the system includes the defining of, and performance of activities in a systematic way from conception, requirements, development or configuration, testing, release and operational use.

The four GAMP Life-cycle phases include:

> Concept
> Planning and Project stage
> Operation
> Retirement

The Concept Stage is concerned with understanding the need or the problem to be addressed. We will see that the User Requirement Specification (along with other specifications) and the initial risk assessment help to drive a project forward in a systematic manner. The most common life-cycle approach for Computerised and Automated systems is the V-Model. The GAMP based V-model lays out a roadmap which facilitates the Validation of equipment and automated systems.

The Planning and Project stage involves the planning of the validation effort required to implement the system into the business area(s) based on identification and approval of system concept. This phase includes assessments of the regulatory and system risks, supplier assessment, development of validation strategies, identification of deliverables that will be generated, definition of the business process the system will support as well as the user requirements which the system will fulfil.

Design & Development and configuration of the hardware and software is also required to meet the system requirements as per specifications. In case of custom Software components, this effort could also include detailed Software design and developmental testing to ensure readiness for verification testing.

Verification – This effort confirms that specifications have been met and releases the system for use. This phase will involve multiple stages of reviews and testing depending on the system type, the development method applied and its use. Once verification activities have begun any changes to the system must be captured through change control.

On successful completion of the verification activities, the system is then released for effective use. The Test strategy and other verification activities will vary widely between simple equipment and more complex customised/ configurable systems. The verification and validation approach is typically agreed and detailed in at the validation planning stage. The VP can be updated accordingly as the project develops with more detail been added. Alternatively, a test strategy document or matrix could be written to provide more specific test plans.

Verification deliverables vary based on the complexity and level or customisation of the system in question. Corporate or company specific procedures also shape the required activities to be completed and reported. Some generic deliverables are listed below.

> Approval, executing and review of test protocols
> Writing and approving SOPs for operation and maintenance of the system
> Traceability Matrix
> Completion of any Risk mitigations (e.g. updates to FMEA etc.)
> Validation Summary Report(s)

Validation reporting requirements varies depending upon the scope of the system and should also be driven by a procedure and template. The Validation Plan can also outline the deliverables and what needs to be addressed in the report. A Validation Summary Report (VSR) shall be written which summarizes the results of executing the VP the documents created for the validation activities, summarizes (or points to summaries) of the testing performed. Finally, the VSR indicate the acceptance of the system/equipment by the user by the Project team and state that the equipment is released for commercial operation / production.

The operation phase supports the need to maintain compliance and fitness for intended use after the system is released for normal use. It is important to ensure the system remains within a continued validated state. All proposed or necessary changes to the system must be assessed and controlled as part of a change control process. Once the system has been accepted and released for use, the operation phase begins. This phase consists of maintaining the system's compliant state and fitness for intended use through the control of the procedures supporting the system's operational use.

During the operation phase the below activities are typically completed:

- Ongoing Training
- Preventative Maintenance
- Service management and performance monitoring.
- Change Control
- Periodic review
- Maintaining system security
- Records management
- Calibration

The retirement phase involves the planning and proper management of activities relating to the removal of systems from service (shutdown). The retirement should take into account the storage of any data and any data migration that needs to occur prior to retirement. The retirement plan, if needed, will outline the retirement strategy from the roles and activities that will be conducted to the removal of the system for use. A Retirement Summary Report is produced that documents the results of the activities defined in the retirement plan including:

- Retirement Plan and Timelines.
- Summaries of any data migration activities.
- Identification of the storage location of documentation relating to the system.
- Obsoleting of SOPs.

It must be stressed that GAMP is a set of principles, a set of guidelines that aim to achieve compliant computerized systems that are fit for intended use. GAMP Guidelines differ to 21 CFR QSR regulations as they are not legal or statutory requirements. However, they represent industry best practice and compliment the Validation efforts that are legal requirements and statutory requirements.

Regulatory Review

Software Validation is a requirement of the Quality System regulation, 21 Code of Federal Regulations (CFR) Part 820. Validation requirements apply to:

(1) software used as components in medical devices,
(2) software that is itself a medical device, and
(3) software used in production of the device or in implementation of the device manufacturer's quality system.

Note: EU GMP Annex 11, provides information on the inspection of 'Computerised Systems'.

In addition, computer systems used to create, modify, and maintain electronic records and to manage electronic signatures are also subject to the validation requirements. Such computer systems must be validated to ensure accuracy, reliability, consistent intended performance, and the ability to discern invalid or altered records. The regulated user should be able to demonstrate through the validation evidence that they have a high level of confidence in the integrity of both the processes executed within the controlling computer system and in those processes controlled by the computer system within the prescribed operating environment.

Specification Hierarchy

An equipment URS can define the requirements of a computerised or automated system along with the operational, functional, process and safety mandated by the customer. If the system is bespoke and complex, an SRS may be written to more clearly detail the software requirements, automation and functionality. Similarly, an FDS (Functional design specification) can be written to address how mechanical or physical processing occurs.

URS-to-SRS

Scenario: a URS is written to specify the requirements for an automated blister packaging line in a Medical device company.
The URS details the following:

URS R1.0 – The machine shall be capable of operating in various modes to allow the manufacture of product and other debugging activity.

In turn, an SRS can be written to interpret the URS, for example, SRS-the system shall have the following modes:

SRS 1.1 Run empty mode -in this mode the equipment does not accept any new product.
SRS 1.2 Production mode – every station operates within the machine.
SRS 1.3 Bypass mode - where any operation can be disabled.

Another 2 examples of a URS requirement been transposed to an SRS requirement are shown below:

Example 1: URS Requirement

URS R1.0 In the event of E-Stop activation, all sequencers shall be maintained and shall retain the sequence step that they were in at the time of E-Stop activation.

SRS Requirement

SRS 1.0 The lot count and Lot integrity must be maintained after E-stop activation.

Example 2: URS Requirement

URS R1.0 The system shall use password protection
SRS Requirement

SRS 1. 1 Basic machine functionality (Cycle Start/Stop, Fault Reset, Manual Operations) require no security.

SRS 1.2 As required, User IDs will be assigned to security groups for authentication. Authentication will be via Active Directory authentication against domain accounts.

SRS 1.3 An auto-logoff feature shall be incorporated in the design.

Examples of Security requirements:

➢ Three levels of access required, operator, and engineer and maintenance
➢ Engineer - access to all screens, to modify process settings Maintenance - access to functions required to perform machine maintenance activities.
➢ Operator - restricted access, does not have access to change process settings.
➢ Different access levels will require different passwords.
➢ No security will be required for basic operations (Start/Stop)
➢ A user auto-logoff feature shall be incorporated in the design. The auto- logoff time shall be configurable.
➢ A soft copy of Program settings must be provided with delivery of the equipment.

Examples of Security requirements:

➢ The reject count and yield must be displayed on the HMI screen.
➢ Real-time readings for all critical parameters shall be visible on the HMI Screen.
➢ All Critical parameters shall be adjustable via the HMI Screen.
➢ The status of each door should be visible on the HMI screen.

Examples of EHS requirements:

➢ Activated E-Stops shall be clearly displayed on the HMI screen with a suitable alarm message generated.
➢ Activated E-Stops shall result in no further movement of the system until the E-stop is reset and all alarms are cleared.

System Categorisation

GAMP 5 makes provision for four categories of software in order to distinguish the level of customization/configurability that exists across software's serving different functions.

GAMP Software Category 1, Operating Systems
GAMP Software Category 2, Non configured software
GAMP Software Category 4, Configurable software packages
GAMP Software Category 5, Custom Software

GAMP Software Category 1, Operating Systems

Category 1, operating systems, covers established commercially available operating systems. These are not subject to validation themselves, the name and version of the operating system must, however, be documented and verified during Installation Qualification (IQ). Application software hosted on operating systems need to be validated.

GAMP Software Category 2, Non configured software

Category 3 covers commercially available, standard software packages and "off the- shelf" solutions for certain processes. The configuration of the software packages should be limited to adaptation to the runtime environment (for example network and printer connections) and the configuration of the process parameters. The name and version of the standard software package should be documented and verified in an Installation Qualification (IQ). Special user requirements, such as security, alarms, messages, or algorithms must be documented and verified in an Operational Qualification (OQ).

GAMP Software Category 4, Configurable software packages

GAMP Software Category 4, Configurable Software Packages Category 4 covers configurable software packages that allow special business and manufacturing processes. This involves configuring predefined software modules. These software packages should only be considered as belonging to Category 4 if they are well-known and mature. Normally, a supplier audit is necessary. If this is not available, the software packages should be handled as Category 5. The name, version, and configuration should be documented and verified in an Installation Qualification (IQ). The functions of the software packages should be verified in terms of the user requirements in an Operational Qualification (OQ). The Validation Plan should take into account the lifecycle model and an assessment of suppliers and software packages.

GAMP Software Category 5, Custom Software

GAMP Software Category 5, Custom Software Custom/Bespoke Software (GAMP Software Cat 5) is software that contains custom code designed or modified specifically for a particular customer. As the code is custom it presents a greater risk. This risk must be mitigated with the right approach to the validation.

GAMP Considerations

Correctly assigning a GAMP software category to equipment, a system or process is an important activity that should be completed early-on in the planning stage of a project. There must of some degree of familiarity with the equipment or system. The manufacturer or vendor can be a source of information that may help the designation. In many cases, companies create tools or processes that help determine what GAMP software category applies. These have different names such as questionnaires, screening tools, planning tools etc.

Risk Assessments

A Risk Assessment process should be applied to cGxP computerized systems in order to identify and mitigate potential risks to (1) patient safety, (2) product quality and (3) data integrity. Results identified through a Risk Assessment help to determine the validation strategy, the effort and time required, and allow better targeting of the validation activities to the highest risks.

The Risk Assessment should be revised during the Software Development Lifecycle (SDLC) if the functionality, requirements or intended use of the system changes. The Risk Assessment activity should also be evaluated during system build-up as well as when implementing changes. Risk Assessment tools for cGxP computerized systems are typically completed during the planning stage, specification stage and post qualification if a change or update is required.

Planning Stage

Initial Impact/Risk Assessment – during the planning phase to identify the level of impact and GxP relevance of the system/equipment. (Tools used: High Level Risk Assessment).

Specification Stage

Functional or Quality Risk Assessment – during the specification phase - identify potential risks and possible mitigations to be to be introduced to the process. (Tools used: Quality Risk Matrix, (p)FMEA).

Changes to the system

Impact Assessment of changes – as part of the change control process in the system operational phase. The following diagram defines the Risk Assessment steps within the System Life Cycle (Tools used: Impact assessment checklist, Change control procedures).

Quality Risk Matrix

A QRM is a risk assessment that identifies and manages the risk to patient safety, product quality and data integrity that relate to the systems processes. Risk Scenarios or potential causes should be developed for each identified function or process step and then assessed for the impact on patient safety, product quality or data integrity. Risk mitigations and controls should then be introduced to address both medium and high levels of risk. The QRM requires 3 "assessments" in order to produce an estimation or overall Risk (Low, medium, high)

Assess Likelihood
Assess Detectability
Assess Severity

Traceability Matrix

A Traceability Matrix should be prepared as required in accordance with company and internal policy. It is also recommended by GAMP guidelines, ASTM E2500 and ISPE Risk based approach to Validation. The matrix links the user requirements and specifications to the testing and validation activities. A traceability matrix illustrates that all user requirements are traceable to the verification/validation activity or vendor documents as relevant (FDS if applicable, Design specifications etc.) A simple traceability matrix (TM) format is shown below on the next slide. Generally, individual organisations will have an approved template to work from. However, the URS structure can form the basis of the template, with additional columns added to document the test/verification method, Reference documents (such as FDS' and vendor specifications and design documents)

Configuration Identification

Software and hardware packages should be identified by a unique product identifier and a version number. For the software end-user, the parts of an automated system that are subject to configuration management should be clearly identified. The system should therefore be broken down into configuration items. These should be identified at an early phase of development so that a complete list of configuration items is defined and maintained. The application-specific items should have a unique name or version ID. The depth of detail when specifying the elements is decided by the needs of the system, and the organization developing that system.

Requirements for the User ID and Password

User ID: The user ID of a system should have a minimum length agreed with the customer and should be unique within the system.

Password: A password should always consist of a combination of numeric and alphanumeric characters. When setting up passwords, the number of characters and a period after which a password expires should be stipulated. The structure of the password is normally selected to suit the specific customer. The configuration is described in the section Security Settings of Password Policy.
Criteria for the structure of a password are as follows:
Minimum length of the password
Use of numeric and alphanumeric characters
Case sensitivity

Audit Trail

The audit trail is a control mechanism of a system that allows all data entered or modified to be traced back to the original data. A reliable and secure audit trail is particularly important in conjunction with the creation, change or deletion of GMP relevant electronic records. In this case, the audit trail must archive and document all the changes or actions made along with the date and time. Typical contents of an audit trail must be recorded and describe the procedures "who changed what and when" (old value/new value).

Uninterrupted Power Supply

An uninterruptible power supply (UPS) is a system for buffering the main power supply. If the power supply fails, the battery of the UPS supplies the required power. When the power supply returns, the UPS battery stops supplying power and is recharged. Some UPS systems provide the option of main power supply monitoring in addition to the buffering function. They guarantee an output voltage at all times without interference voltages. UPS systems are necessary so that process and audit trail data can continue to be recorded during power failures. The design of the UPS must be agreed with the system user and must be specified in the URS, FS or DS. The following points must be considered:

➢ Energy requirements of the systems to be supplied
➢ Power of the UPS
➢ Required duration of UPS buffering

The energy requirements of the systems to be buffered decide the size of the UPS. A further selection criterion is the priority of the systems. Systems with high-priority include:

➢ Automation system (AS)
➢ Archive server
➢ Operator station (OS) server
➢ Operator station (OS) clients
➢ Network components

Field devices that generally have relatively high energy requirements may also be included in the buffering depending on the power of the UPS. This must be decided in consultation with the system user and related to the classification of the process. Whatever is decided, it is important that the systems for logging data are included in the buffering. The time at which the power failure occurred should also be recorded. The use of UPS systems involves the installation of software. This should be installed and configured on the PC-based computers of the process control system to be buffered. The setup should also account for:

➢ Configuration of the power failure alarms
➢ Stipulation of the time before the PC is shut down
➢ Stipulation of the time during which UPS buffering is provided

The automation systems (AS) must be programmed so that the process control system changes to a safe state after a selectable buffer time if a power failure occurs.

Types of UPS

Due to the different requirements of the various devices involved, three classes have established themselves as stipulated by the International Engineering Consortium (IEC) in product standard IEC 62040-3 and the European Union EN 50091-3:

Offline UPS

The simplest and least expensive UPS systems (according to IEC 62040-3.2.20, UPS class 3) are standby or offline UPS systems. They protect only against power outages and brief voltage fluctuations and peaks. Undervoltage and overvoltage are not compensated. Offline UPS systems switch to battery supply automatically if there is overvoltage or undervoltage.

Line-interactive UPS

The way in which line-interactive UPS systems (according to IEC 62040-3.2.18, class 2) function is similar to standby UPS systems. They protect against power outage and brief voltage peaks and can compensate voltage fluctuations continuously using filters.
Online UPS

Double conversion or online UPS systems (according to IEC 62040-3.2.16, class 1) count as genuine power generators that continuously generate their own line voltage. Connected consumers are therefore supplied permanently with line power without restrictions. At the same time, the battery is charged

Software-Source Code Review

For GAMP Software Categories 4 and 5 source code review is advised unless the supplier has evidence of the same available for review. As part of Good Automated Manufacturing Practices, reviews should be completed as part of the development lifecycle. If a source code review is not completed a justifiable rationale should be documented in an applicable document such as a Validation Master Plan.

Calibration

A key part of any qualification is to confirm that the equipment is fit for the intended purpose. Each piece of equipment will have a defined operating range. For example an oven may have an operating range of 20°C to 100°C ±5°C. However, the process window may only require a temperature range of 30°C to 60°C. In this instance a calibrated range of 20°C to 70°C would suffice. However, if the process window or the temperatures at which product was manufactured ranged from 20°C to 100°C this would present a problem as it falls outside of the equipment qualification range when the calibration tolerance is taken into account.

Deviations

A deviation can be simply described as an unintended event which causes a test or verification to fail to meet expected acceptance criteria. Each company or organisation should have a procedure detailing the management of deviations. It is critical that all deviations are identified, investigated and evaluated for their impact on product quality, the risk/impact to the patient and the impact on the qualification or validation. The basic components to a deviation are listed below:

> Deviation Description- provides the page and section of the deviation and an overall description eg. document generation error, operator error, machine crash etc.
> Potential impact on product – does the deviation impact the product?

- ➢ Potential impact on validation/qualification-will the validation have to be repeated in part or in full?
- ➢ Investigation- DMAIC, RCA, Fishbone Diagram, 5W
- ➢ Root Cause- what is the concluding root cause?
- ➢ Planned Resolution- what actions are required to be implemented?
- ➢ Deviation Resolution (Actions completed)- were all the actions in the planned resolution implemented? What is the final result? Have the actions been effective? Requalification

Over the lifetime of a piece of equipment, the need to requalify may arise. Therefore, any proposed change to equipment or a process must be assessed to see if the validated state will be impacted. It is therefore critical to understand clearly the nature of the change(s). Some scenarios where requalification of equipment may be required include:

- ➢ Major equipment repairs
- ➢ Moving equipment
- ➢ Changes to the upper and lower operating limits of the equipment
- ➢ Upgrading of software
- ➢ Hardware upgrades or changes
- ➢ Changes in performance and/or defect levels

After assessing any proposed changes based on the reasons listed above a determination of the level of requalification is required. This may be limited to a partial requalification (addendum) or it may require a full requalification.

CHAPTER 9

Process Validation

Introduction

This chapter provides an introduction to Process Validation for medical devices. Process validation is a statutory and regulatory requirement for the manufacture of medical devices. Per FDA 21 Code of Federal Regulations Process Validation is a regulatory requirement of Good Manufacturing Practices (GMP) for both pharmaceuticals (21 CFR 211) and medical devices (21 CFR 820). In addition to the regulatory drivers, process validation is a requirement in order to obtain certification to international standards issued by many notified bodies. (e.g. ISO 13485 Medical Devices- Quality Management Systems, ASTM E2500-Standard Guide for Specification, Design, and Verification of Pharmaceutical and Biopharmaceutical Manufacturing Systems and Equipment etc.)

Traditional and New Approaches to Validation

Historically, process validation involved the testing and verification of all aspects of a process. While this may seem appropriate, it must be understood that in order to test/verify all aspects of a process, for it to hold weight, this activity must be documented and recorded. In this respect, an "all aspects" approach to process validation can be burdensome to resources. The traditional approach largely used the V-Model which set out a sequence of deliverables that should be completed. The use of risk assessments were limited as all requirements of a system were tested and qualified.

In recent years, a risk based approach has been increasingly endorsed by regulatory authorities and hence adopted by medical device manufacturers. One such standard is the ASTM E2500. As the title suggests, it is primarily used within pharmaceutical and biopharmaceutical industries, its principles and core approach can be adopted by medical device manufacturers also. ASTM E2500 was designed to make the implementation process for GMP systems and validation more cost-effective. It aims to achieve this based on scientific and risk-based principles, focusing on the risk to the patient. However, at just a five-page document, ASTM E2500 lacks the detail required in order to meet regulatory expectations. While different terminology and philosophies exist they do not change the regulatory expectations relating to validation. Both approaches exhibit common elements which include:

➢ Good engineering practices
➢ Planning
➢ Requirements definition (URS etc.)
➢ Design review
➢ Change Management
➢ Documented testing and inspection

While many manufacturers may predominantly choose a particular approach, it is common to see elements of both approaches (traditional & risk based). Each individual company will shape its internal validation procedures to best suit its business needs of the company.

What Is Process- Operational Qualification (OQ-P)?

The ability of a process to produce product in accordance with pre-determined specifications under worst case conditions. PQ is only required if no worst case conditions are evident.

What Is Process-Performance Qualification (PQ)?

The ability of a process to consistently produce product in accordance with pre-determined specifications under anticipated conditions (normal/routine conditions). Before considering Process Validation in further detail, it is important to look at the pre-requisites and other supporting activities required. These are examined in the sections below.

Test Methods & Process Validation

It is important to consider test methods early on in the validation lifecycle. Before you can begin to consider Process Validation, test methods should be understood and in place.

A Test Method is a process or an action used to verify that a product feature meets a predefined specification. Tests methods can be physical or analytical in nature. Test Method validation should be completed in advance of process validations to allow the proper assessment of process and product outputs meaning it is often a pre-requisite to Process Validation.

Examples of test methods include simple visual inspection by microscope, measurement of a dimension with a calipers or measurement of a dimension using an automated optical inspection system. Some test methods will involve MSA (Measurement System Analysis) studies for example, a measurement of a dimension by an operator using a microscope. In contrast, a test method to determine organic residuals would require an Analytical Test Method validation.

The equipment must be qualified (Installation Qualification and Operational Qualification) before the method is validated. Remember – Testing completed in contract laboratories or specialist services also require validation! Test methods are critical to the success and integrity of your Process Validation as they assess the outputs. E.g. what are the dimensions, physical attributes or chemical properties of the product and how do they confirm to specifications?

Stages of Process Validation

As outlined in previous chapters, the 3 stages of Process Validation include:

➢ Process Design
➢ Process Validation
➢ Continued Process Monitoring

During the Process Design stage, the Commercial Manufacturing process must be established during the process design phase. Some typical activities include:

➢ Definition of Process Inputs

> ➢ Effects of inputs
> ➢ Process Outputs – CQAs (Critical Quality Attributes)
> ➢ Establishing process windows
> ➢ DFMEA /PFMEA (Design/Process)

Design Control Procedures should be developed to allow proper management of the Process Design Stage. At the Process Design stage, The Business must define the manufacturing process. This often involves liaising with vendors and Subject Matter Experts (SMEs). Process Qualification Stage looks at the validation of process design to confirm process is operating as intended and is capable of consistently producing product to meet quality requirements. Finally, stage 3, Continued Process Verification provides ongoing assurance through regular testing and verification to ensure the process is in control. Stage 3 is often referred to as In-Process Control or In-Process Testing (IPC/IPT). This data provides feedback to engineers allowing them to trend the performance of output data. This can identify deficient equipment, changes in wear tooling etc.

Fundamentals of Process Validation

The most important point when it comes to validation is that validation is neither exploratory nor investigative. Equally, it is not an engineering study. If you are ready to validate a system or process, all of the groundwork must be completed. This means critical parameters must be defined and documented, with technical rationale on why such parameters are critical etc. This body of work is typically done during a process development study or protocol. Validation of confirmation, so Process Validation is confirming that a process is capable of consistently manufacturing product under anticipated conditions. Remember, validation should be representative of the commercial process, so any issues in Process Validation will be repeated in commercial manufacturing.

Consistency, a core principle of Process Validation is typically demonstrated by producing 3 batches/runs for a Process Performance Qualification (PPQ). These batches should be representative of normal production i.e. the size of the batch should be typical of commercial volumes. The PQ study should be executed at nominal conditions, (often termed "anticipated conditions") essentially referring to a controlled environment. Controlled material and controlled parameters (CPPs) are required. Nominal settings should be selected for PQ.

Process Validation and Dominance Factors

The concept of dominance is a term used to describe the "influential" or "dominating" effect on a system or process. Typical examples include the injection moulding process, and packaging process. For example, an injection moulding process can be said to have material as a dominant factor. Batch to batch differences of resin or raw material may cause a change to outputs such as dimensions of a product or component. If dominant factors cannot be identified or understood a "Designed Experiment" (DoE) technique can be used to properly determine them. Dominance can be categorised into 5 sections. (1) Setup Dominance (2) Time Dominance (3) Worker Dominance (4) Information Dominance and (5) Component Dominance.

Setup Dominance

Setup Dominance -The Process or Equipment relies principally on a procedure or process setup. Process should be stable one "set-up".

Examples include ovens and package sealers. With regard to the oven, the setup would generally be controlled by a recipe or program. This program would be selected by the operator through the Human Machine Interface (HMI). The setup with the correct version of the recipe that contains the desired temperatures, times and pressures is therefore a critical input to the process. With regard to the Packaging Machine (blister packaging), the correct setup for the tooling and program are critical inputs. If Setup Dominance is significant, it is best practice to have 3 separate set-ups/changeovers in the Performance Qualification (PQ).

Time Dominance

The Process or Equipment is subject to changes over time (drift over time in temperature, solvent cleanliness, tool wear etc.). The process may need a schedule of process checks and adjustments to ensure process consistency.

Examples include CNC Machinery (tool wear) or aqueous based cleaning systems. The tool may only be able to manufacture 1000 parts before defects or quality issues are encountered. If Time Dominance is significant 3 time points or cycles of expected variation should be made e.g. 3 points in the cycle (start, middle and end) or 3 points in a shift (start of shift, middle of shift and end of shift).

Worker Dominance

For Worker dominance, the process requires operator experience and skill. Examples include manual or hand finishing. If Dominance is significant, ensure there are a minimum of 3 operators involved in the manufacturing/ activity.

Information Dominance

With information dominance, the process or Equipment requires the transmission and/or analysis of information. Examples include LIMS, MRP and ERP systems. A minimum of 3 information transmissions in the PQ should be completed.

Component Dominance

The Process is influenced by the variability of the input materials and/or components. It requires robust inspection and sorting procedures as well as process adjustments. When Component Dominance is significant, ensure there are a minimum of 3 component/raw material batches in the PQ sampling plan. If component dominance is significant this can be mitigated by including the material/component variation in "worst case" testing as part of the Operational Qualification Process (OQ-P)

Process Operational Qualification (OQ-P)

During the Operational Qualification-Process (OQ-P) study worst-case process conditions are normally employed. This may be worst case temperatures, speeds, feeds etc. The OQ-P should challenge the manufacture/processing of product at the limits of the processing window (process range). If no worst-case conditions exist, then an OQ may not be required and only a Performance Qualification is required.

A family or matrix approach is often used where similar products are to be validated. A particular product size of product configuration may be selected to represent the worst-case product. Therefore, by qualifying the worst case, all other products within that family of products would be considered validated. However this approach must be clearly documented and technical rationale provided in advance of any qualification activities. This can be addressed in a Validation Plan or within a protocol.

Protocol Approval Check list

The Validation Protocol is the means by which objective evidence is documented and gathered. The Validation Protocol is therefore a critical document. It should clearly set out the approach to the validation, detailing methods, tests and verifications to be completed and the acceptance criteria that applies to such tests and verifications. Remember, a validation document is a legal and regulatory document and can be subject to detailed scrutiny. Below are some suggested general checks to apply when writing Validation Protocols.

Author
- SOP available -Protocol conforms to validation procedure.
 - Ensure item numbers and batch size are correct.
 -Test methods are correct.

SME Reviewer
- Is the protocol number correct?
-Review content of Protocol for accuracy and completeness.
 - Protocol conforms to validation procedures.
 - Procedure and evaluation table are appropriate and correct.   Engineering:
 - Review content of Protocol for accuracy and completeness.
 - Specifications and operating parameters are correct.

QC / Laboratory
- Review content of protocol.
 - Raw Material Specifications are in place.
 - Finished Product Specifications are in place.
 - Testing and sample size is correct.

Quality
- Review content of protocol.
 - Protocol conforms to SOPs.

- Evaluation and acceptance criteria are appropriate.

Process Performance Qualification

The purpose of the PPQ is to demonstrate the capability of the process to consistently manufacture product to pre-determined specifications under normal operating conditions and defined parameters. Validation is confirmation, so process validation is confirming that a process is capable of consistently manufacturing product under anticipated conditions.

- ➢ Lots should be produced consecutively (in sequence)
- ➢ Lots must meet the acceptance criteria set out in the protocol
- ➢ The lot size should be reflective of the intended lot size and also take into account normal variation
- ➢ If a family approach or matrix approach is used, the product selection must be clearly justified and documented
- ➢ Execute under anticipated conditions, essentially this refers to a controlled environment. Controlled material, controlled parameters (CPPs)
- ➢ Nominal settings should be selected for PPQ

Yield Data (aka Process Yield Data)

Process Yield is a term used in manufacturing to represent the overall process performance. Yield is most often expressed as a percentage of good/passing product. It reports the % of compliant units, that is units or products that meet the product acceptance criteria (eg. CQAs). The remaining "bad" units are classified as defects or scrap. In some manufacturing processes, rework is possible or permitted.

Yield data often forms part of the acceptance criteria for a validation. The overall process yield for each batch should be calculated and compared to the starting process weights or units to determine loss due to processing as it is common to lose material during processing.

Continued Process Verification

Once the initial validation is completed it is important that the system or process remains within the validated state, meaning that the system remains in a state of controlling process systems that capture information and data about the performance of the process. The use of statistical trending techniques should be considered. Data analysis of process and product should also include trending of raw materials, components and finished product. The purpose of process monitoring is to ensure critical parameters remain within control limits. It also helps to identify increasing variability or instability within the process which can then be investigated. All processes must have an upper and lower limit. If a process parameter only has a one-sided limit, then provide rationale in the OQ protocol to justify why a one-sided parameter window is acceptable. This requirement is not applicable to parameters that are set points.

Revalidation (or Maintaining a Validated State)

Revalidation is sometimes required if the original validation is no longer valid or representative of the process. Some instances where revalidation must be considered include changes to the process that can affect the product quality or efficacy, a removal or the addition of a processing step or transfer of the equipment to a different location. In many companies an Impact Assessment is conducted if there is a proposal to modify a manufacturing process. Some changes may not require any validation while others may require a verification run.

When changes are proposed to the validated state of a process, the proposed changes must be fully understood in terms of the impact to product quality and the validated state. A risk assessment should be conducted to determine risks and appropriate mitigations.

Scenarios on maintaining a Validation State

Line Addition-Product (New Product or Product Transfer)

This may be required if a new product has been introduced but uses the same process(es) for manufacturing. Typically this can apply if a new size has been introduced. For example, a new size of surgical blade.

Line Extension

A line extension commonly refers to a scenario where the product/process is different or considered outside the existing range or processing parameters.
Note: The impact on the validated state for line additions and line extensions should be assessed formally and documented. A line extension may require a new validation or addendum to the existing validation, whereas a line addition to add a new product may be within the scope of existing validations.

Acceptance Criteria

The acceptance criteria contained in Validation Protocols are normally based on established product specifications. For example, a contact lens manufacturing company may produce a lens with a diameter of 18mm ±0.2mm. The product produced during a process validation must be inspected to record the diameters of lenses being manufactured. Disposition of product is based on the product specification and determines if the product feature measured receives a pass or fail.
In addition to Product Specifications, it is common to have acceptance criteria such as Yield, and OEE. The acceptance criteria for these conditions are normally driven by an internal company procedure or alternatively can be detailed in the validation plan or protocol study.

Validation Strategies

A Family Approach (a.k.a. Bracketing, Matrix Approach) to validation is often used where a variety of similar products are manufactured using the same equipment. For a Process Validation, a product that is representative of the family or group of products may be selected. Alternatively, a 'worst case' product may be selected as it presents the greatest challenge to manufacture to product specifications.

Principles of Worst Case Selection

Worst Case is a particular condition, set of conditions, and/or set of process parameters, generally made up of processing limits. Worst case conditions present the greatest chance of process issues or the greatest chance of failures due to product quality. Worst case conditions are used at OQ-P stage to provide the greatest level of challenge, however, this is outside of normal operating conditions.

Requalification

During the lifetime or a process or piece of equipment, the need to re-qualify may arise. Such need should be assessed according to a validation procedure. Generally, the same tools used in the original validation can be re-applied to identify the need or re-qualify and indicate what requirements must be included.

The first step must be a review of the existing qualification, as changes may not impact the validated state, or may only require a limited requalification. For example, moving a piece of equipment may only require requalification of the utilities such as compressed air or process water if the operation of the equipment is not impacted by the movement and re-siting.
Some examples where re-qualification may be required include:

> Transferring a process from one plant to another plant
> Changes to the process settings which may impact the product quality
> Changes to the design of the product
> Changes to manufacturing aids (e.g. cleaning agents, jigs and fixtures)

Examples of Process Validation Deficiencies

Identification of Critical Process Parameters

No procedure in place to define or identify Critical Process Parameters (CPP). Not documenting from where these CPPs would be obtained when writing Process

Validation Protocols

In a drying process an oven, no rationale was documented as to why time and pressure were not considered CPPs. Temperature was only considered a CPP.

<u>Reproducibility</u>

For a concurrent validation, three separate reports were drawn up and each batch had been concurrently released. However, no summary report was generated to assess the reproducibility of the process.

<u>Risk Assessment</u>

No linkage between the Validation Protocol and the various controls in the manufacturing process that had been identified in the risk assessment as important for risk mitigation.

Case Studies on Process Validation

The following case studies review key considerations when it comes to Process Validation. For your benefit, we have focused on the critical elements which include:

1) General Principles of Process Validation
2) Dominance Factors
3) Parameters and Settings
4) Other Considerations

Case Study -CNC Grinder –Performance Qualification (PQ)
System Overview

A Computer Numerically Controlled (CNC) Grinder uses grinding wheels to machine complex geometries or modify surfaces. As with any Process Validation, it should factor in when to influence the process. (Dominant Factors). The machine settings are also another factor which needs to be considered for Process Validation. Machine settings should also have a tolerance or +/- value associated with them. However, with CNC machines, as they are numerically controlled, parameters such as spindle speed tend to be accurate. In the other case studies, we will see other parameters such as temperature and how tolerances are associated with settings.

<u>General Principles of Process Validation applied to a CNC Grinder</u>

As the spindle speed of the machine is "set-point" only and is known to be accurate, an Operational Qualification (OQ) is not required as there are no 'Worst Case' settings. All system settings are set-points and the only variable is 'wheel life'. As the wheel is used to manufacture/grind parts, over time the wheels grinding performance decreases, therefore, the wheel has a life expectancy. Wheel life assessment should form part of the Performance Qualification. A Performance Qualification manufactures components at start, then middle and at the end of wheel life. First Article inspection of (FAI) should be performed at the start of each run in order to confirm the "first off" is according to specification and that the machine setup is correct.

<u>Dominance Factors</u>

For a CNC grinder, appropriate dominant factors include time and setup. Time is a factor as the wheel life changes over time. Setup is also a factor as the tooling and fixturing must be setup accordingly and this can be a source of variation that must be challenged as part of the PQ.

Parameters and Settings

Coolant temperature, grinding feed rate, cutting dwell time and spindle speed all have the ability to affect performance and product quality, therefore they can be considered controlled parameters. Coolant temperature should have a tolerance specified in the validation protocol e.g. +/- 2 degrees.

Other Considerations

The incoming raw materials e.g. mild steel bar stock, should confirm to an acceptance criterion of Certificate of Analysis.

Case Study – Cleanline Operational Qualification and Performance Qualification (OQ-P/PQ)

System Overview

Cleanlines are used to clean fixtures, equipment parts and products within the medical device industry. Cleaning processes can be loosely divided into intermediate cleaning processes and final cleaning processes. A higher acceptance criterion is applied to final cleaning processes. Cleanlines are aqueous or solvent based systems. Aqueous systems typically use deionised water and detergent, followed by DI rinsing and dryings steps. Solvent systems are also effective at removing grease and oils from parts and components.

General Principles of Process Validation Applied to a Solvent Based Cleanline

A cleaning process is made up of 1) cleaning parameters such as temperature, time and ultrasonics 2) manufacturing agent, in this case-solvent and 3) worst case conditions. The heating or cooling of liquids (although temperature controlled by a PLC-temperature probe) is subject to drift. This drift can vary on a given day, or due to the use of the equipment. Therefore, it is important to quality check an operating range for the equipment and the cleaning process. This operating range is qualified by the execution of an OQ-P, Operational Qualification-Process. The OQ-P represents the limits at which products will be cleaned. The OQ-P aims to demonstrate that if the process settings are subject to drift in time or temperature, the cleaning process is still effective as it operates within the validated operating range. This operating range is also known as the process window.

The following strategy should be applied for the Operational Qualification- Process and Performance Qualification of a cleaning system:
OQ Low, 1 lot
OQ High, 1 lot
PQ nominal, 3 lots

PQ nominal Rework, 1 lot

Full loading conditions should be applied for all cleaning run/conditions. This means that the basket or "carrier" that holds the pieces while in the washer, is full to the normal, anticipated capacity. This is to create the conditions as they are intended during commercial manufacturing.

Dominance Factors

The Dominant Factors for a cleaning process are considered to be time based. As the system is used, the water/solvent in the tank gets more "soiled" as time moves on. It essentially gets dirtier; therefore, there may be a difference in the cleaning performance of the system at the start of the day (clean water with fresh detergent) opposed to the end of the day, after a full shift of cycles (dirtier conditions).

Parameters and Settings

Parameters include rinse temperatures, times and ultrasonic frequencies. Some equipment may be fitted with in-line conductivity meters and/or pH probes which indicate the "cleanliness" of the solution. If a drying stage is incorporated into the system, this will also have temperature and time settings and tolerances. Note, that some basic systems may only have fixed ultrasonics which cannot be varied.

Other Considerations

An important consideration when conducting cleaning Process Validation for medical devices is the manufacturing agents (greases, oils etc.) that are used up-stream in intermediate steps or processes. The chosen cleaning detergent or cleaning solvent should be selected in a systematic way to ensure it is effective in removing the "soiled" material. Some key points to assist in this process are:

Map out the complete manufacturing process identifying the stage at which cleaning is completed.

CHAPTER 11

Test Method Validation

Introduction

This chapter covers the design, execution and analysis of test method validation for medical devices. Test method validation involves the formal documentation of a test method used to capture and analyse data or information. The reason test methods need to be validated is to confirm that they are suitable and fit for the intended purpose. Secondly, and of equal importance is the need to verify that the test method performs to an acceptable level and is reliable and trustworthy over time. After all, test methods are used to assess product outputs such as dimensions, material strength and product functions. Getting this wrong will lead nowhere very quickly, so it is important to have confidence in the results of testing.

Validation studies must demonstrate method capabilities in the testing environment. As a result, validation studies allow the formal documentation of the ruggedness of the test method in real-use conditions (i.e. demonstrating that the precision and accuracy limits are met with different technicians, different production batches and variable test equipment, etc.).

The form and shape of any test method validation is influenced by several factors namely (1) in house requirements, these are internal procedures within a company that define the process and may provide or require the use of company templates or forms etc. (2) external requirements such as those relating to ISO and ASTM standards, (3) the type of test to be completed – visual inspection, manual contact measurement, non-contact (automated) measurement system, destructive tests and so on, (4) the type of equipment and status of equipment- simple equipment may not require equipment validation. However, complex equipment such as equipment that uses automation and software needs to be fully qualified. In addition engineering studies may be require to test the robustness of the system, prior to measurement capability studies or formal test method validation. Also, existing equipment may be suitable for use and already qualified. Therefore an assessment of the suitability for use must be completed to determine if all requirements are met. (5) Product range and specifications. If a family of product are to be tested, the test method validation can be more complex. The product specifications is a critical input as the accuracy of equipment needs to be factored in, and this is dependent on the type of measurements required to be taken, and the upper and lower tolerances to be applied.

Examples of Test Method Validations

Example 1

A packaging company has a seal strength on the lid of a package. It wants to put in place a test method to test the seal strength of the package. This scenario would call for a test method validation.

Example 2

A medical device incorporates the use of a spring that is used to actuate a valve. The manufacturer of the device wants to develop a test method that examines the tensile strength of the spring on an ongoing basis. This scenario would also call for a test method validation.

Example 3

A contact lens manufacturer uses an optical comparator to measure the diameter of contact lenses during manufacture. The manufacturer must develop and validate a test method to facilitate the measuring of contact lenses.

Changes Requiring Re-Validation

Take the case where a standard test method is established and in operation. However, a change to the system software is required. This type of change could impact the measured output. Therefore the change needs to be considered for re-validation.

Any other changes to the test procedure such as a change in handling of test specimens or the change, addition, removal or modification of equipment including fixturing can impact the measured output.

It is important to note that validation of a test method is not required on each individual piece of equipment or fixturing, once replicate equipment or fixturing is assessed during the validation study. Some examples of changes not necessarily requiring any re-validation or change to a validation report etc. include:

- Clerical corrections to the test method that do not change the method or affect the measurement of the output.

- Removing of referenced supplies that do not impact test output, for example lint or cleaning agents.

- Movement of equipment does normally not merit re-validation of the test method, but a limited equipment qualification may be required.

Test method validations should be product and site specific. This means the site and product should be clearly defined in the scope of the test method validation documents. Before an already existing validated test method can be used with a new product or at a new site, the suitability of the existing test method must be documented. Suitability reports are examined further along in this book.

Definitions

Attribute: is defined as the result of a property or characteristic. It is generally used with the terms pass or fail.

Accuracy: can also be defined as trueness. An expression of the closeness of agreement between the value that is accepted, either as a conventional true value or an accepted reference value and the value obtained. A system with low bias implies good accuracy and vice versa.

ANOVA (Analysis of Variance): a statistical method used to evaluate the significance of differences in means due to different factor-level combinations.

Bias: The difference between observed "average of measurements" and a reference value; also referred to as accuracy.

CQA (Critical-to-Quality): a property or characteristic with specific nominal value and appropriate limit and range providing a particular quality attribute.

Critical Process Parameter (CPP): a process parameter that has a direct impact on critical quality attributes.

Dichotomous Variable: an output with only two possible values. Also known as dummy or indicator variable.

Equipment Qualification: establishing documented evidence that the process equipment is suitable for the intended use and is capable of consistently operating within established limits and tolerances under normal operating conditions.

Process Validation: process validation is defined as confirmation via documented evidence that a particular process performs consistently to a high degree of assurance in accordance with predetermined specifications under anticipated conditions.

Measurement Capability Index (MCI): the Measurement Capability Index (MCI) represents the capability of the measurement system. It is used to evaluate the capability of the gauge to classify product against predetermined specifications.

MSA: a study to determine the degree of error involved in measuring the given parameter. The measurement system involves the combination of operations, procedures, gauges, instruments, environmental conditions, people and software.

Precision: the degree of agreement (scatter) between a series of measurements when a method is applied repeatedly to multiple samplings of a homogeneous sample or artificially prepared sample under the prescribed conditions. There are three types of precision; repeatability, intermediate precision and reproducibility.

Range: range is defined as the interval between the upper and lower measurements required. The minimum specified range should be within the equipment range and validated to operate at all points within the range.

Ruggedness (Intermediate Precision): variation on different days or with different analysts and equipment. The extent to which intermediate precision should be established depends on the circumstances under which the method is intended to be used.

Resolution: the smallest unit of measure that can be obtained reliably from a measurement device, also known as gauge discrimination.

Gauge R&R: represents the estimate of the measurement variation. The measurement variation has two components; repeatability or the precision under the same operating conditions (same operator, test method, sample, etc.) and reproducibility or the precision between operators when measuring the same sample with the same gauge.

Variable: is generally the output that is measured.

Validation: confirmation by examination and provision of objective evidence that the particular requirements for a specific intended use can be consistently fulfilled.

New Test Methods

A test method procedure should be created as early on as possible and trialed and examined for completeness and appropriateness. If new test methods are required, a revision controlled draft should be available for the purposes of the test method validation.

Method Transfer

If an existing test method is suitable for the test method validation, a suitability report can be completed to document the suitability and show that all factors have been considered (see attachment 1). However, the test method should have been previously validated. The parameters at which the validation is to be conducted must be within the existing validated range.

Equipment used in a test method must be assessed to ensure the process is within the equipment qualification. All validation testing must be done on qualified equipment. Equipment qualification is therefore a prerequisite of test method validation.

Test Method Ruggedness Study Protocols

Ruggedness refers to the variation, on different days or with different operators or equipment. The extent to which ruggedness (aka intermediate precision) should be established depends on the circumstances under which the method is intended to be used.

An initial ruggedness assessment should be completed to understand the sources of variation. More formal ruggedness studies may be required which should be captured in a formal study protocol. The output of any ruggedness studies should detail any changes or modifications to the test method procedure. Generally, a scoring system is used to describe ruggedness which forms a ruggedness assessment. As a result of ruggedness studies and consequent updates to the procedure, the ruggedness assessment needs to be reassessed. This reassessment will be reflected in the final scores of the Ruggedness Assessment Matrix.

Key Concepts

Accuracy

Accuracy is a measure of exactness of the test method output or another way of putting it is the closeness of agreement between a set of test results. For example, take a component that weighs exactly 4 kg according to an NIST traceable scale. If the weight of component is taken 10 times on the balance under study using the test method under study then calculate the mean weight of the 10 readings. The offset between the mean weight and the 4kg "accepted reference value" is a measure of bias. A large bias = poor accuracy. A small bias = good accuracy.

It is important to note that accuracy does not address the variation between individual measurements. Simply put, if the average is very close to 4kg, then the test method could have been declared to be very accurate. It is advised that you consult any relevant standards (e.g. ISO, ASTM) to the product or feature being measured as standards often will call out an accuracy requirement. Generally, results should be accurate to $\pm 1\%$ of the measured value. Therefore, the equipment must be fit for the intended purpose or the measurements in mind. Note: instrument or equipment accuracy can normally be found on calibration certs provided by the manufacturer or vendor.

Precision

The precision of a method is the degree of agreement among individual test results when the same test method or procedure is applied repeatedly to multiple samplings that represent a population. Precision can be a measure of either the degree of reproducibility or of repeatability of the method. Repeatability refers to the use of a method using the same operator/test person with the same equipment. Repeatability should be assessed using either a minimum of 9 determinations covering the specified range for the method (e.g. 3 concentrations/3 replicates each). Reproducibility refers to the use of the analytical method in different laboratories such as in a collaborative study.

Ruggedness

Intermediate precision (also known as ruggedness) expresses differences related to laboratory variation, as on different days, or with different analysts or equipment within the same laboratory. The extent to which intermediate precision should be established depends on the circumstances under which the method is intended to be used. The effects of random events on the precision of the analytical method should be established. The use of experimental design (matrix) may be used to study the effects of typical variation (dominance factors) on the analytical method (e.g. equipment, analyst, days).

Representative/Continuous Sampling

Representative sampling is used to determine overall process performance (e.g. Pp / Ppk), which is more applicable for processes known or suspected as less than stable or not in statistical control. Sampling in this way best determines overall spread, which includes within-time and time-to-time variation.

Below, some examples are given on how to sample representatively:

Sampling over a given time-period: e.g. a tray of product is produced every 15 minutes, the period of interest is a 1 hour interval and the sample size is 40.

Sampling a batch or product lot not assembled in any order: if the product is packed in a tray (without any grouping) then sample from various sections of the tray.

Consecutive Sampling

This type of sampling involves taking one sample immediately after each another for the subgroup or time period in question, and is used to determine process capability (e.g. Cp / Cpk). Consecutive sampling is used in particular to create control charts where a process is sampled in time order by selecting a subgroup sample consecutively and repeating this sampling over a number of subgroups while in same time order. This method is typically used when the process is stable as there will be little or no causes of lot-to-lot variation.

Range

The range is defined as the interval between the upper and lower measurements required. The minimum specified range should be within the equipment range and validated to operate at all points within the range. If an existing test method or piece of equipment is to be used, it is important to determine if the method parameters for the new/modified test method are within the validated range of the equipment qualification. Remember, all validation testing must be done on qualified equipment. Typically, the equipment qualification assessment is documented in the test method validation protocol.

Resolution

We have previously defined resolution as the smallest unit of measure that can be obtained reliably from a measurement device or system. For example, a Vernier callipers may have different models with different resolutions. Some will have only two digits to the right of the decimal point (X.XX mm) and other models could read three digits to the right of the decimal point (X.XXX mm). The instrument resolution should be better than the resolution of the product specification. If the product specification is X.XXX, then at least a "four-digit" measurement device should be used.

Probability of False Alarms P (FA)

This signifies the likelihood of rejecting a conforming unit. This is typically an acceptance criterion for attribute tests. Refer to MSA template for further illustration.

Probability of Misses P (M)

This indicates the likelihood of accepting a non-conforming unit. This also is typically an acceptance criterion for attribute tests. Refer to MSA template for further illustration.

Test Method Validation Protocols

Typically, an approved template is used to create a validation protocol. The protocol sets out the approach to the validation i.e. the approach to qualify the test method. Refer to the appendix for an example of a validation protocol template.

Attachments to the protocol should include ruggedness assessments completed and references to supporting studies/reports. The drafted or "redlined" test method should be attached to the protocol also. The type of MSA protocol (attribute or variable) should also be determined in the validation protocol.

Test Method Accuracy

Accuracy is influenced by both the instrument (scale) and the test method. If you drop the object on the scale and take a reading before the scale has stabilised, the accuracy is likely to be poorer than when using a test method that demands allowing the scale to stabilise.

Examples include: Tensile strength at break - strength does not exist as a material property independent of the test method used to measure it. For properties like time, distance, and mass, there are NIST traceable standards that can be measured. These standards have a generally accepted reference value that can be compared to the observed readings to assess accuracy (bias). No such reference sample exists for tensile strength at break, deflation time or implant radial strength. For tests without a reference value, the accuracy of the underlying sensor (e.g. load cell) used to determine the output should be addressed if possible.

Components of Test Method Validation

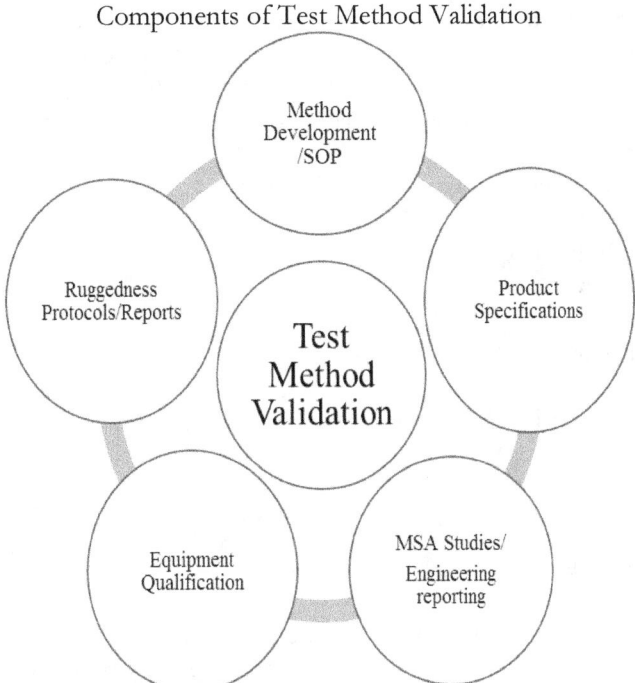

Figure 16: Typical components of a test method validation.

A test method validation ties together many different activities and documents to ensure the method is fit for purpose. The test method itself, (or SOP) must be suitable for the equipment or test system that is to be used. Equipment and tooling(if required) must meet the anticipated range of measurements likely to be recorded during testing. Equipment must also be calibrated and validated accordingly. The establishment of a Product specification also needs to be understood and is a key part of the scope and inputs to a test method validation. The product range to be tested along with the product specifications and associated tolerances feed into standard operating procedures and the test method validation. Ruggedness reports to assess intermediate precision may be required to determine the extent of random effects on the test method. MSA studies are sometimes completed to assess the measurement system in terms of repeatability and reproducibility. The capability of the measurement system is expressed in % terms. Typically a result of 33% or above is required to accept the system as effective at measuring the intended CQA's.

MSA Studies

A measurement system analysis (MSA) is an experimental design used to identify the elements that affect measurement variation. There are two types of data in which MSA studies can be completed i.e. variable data and attribute data. These terms are defined below.

Variable data: data that can assume a range of numerical responses on a continuous scale. Most measurements yield variable data.
Attribute data: data that represents the absence or presence of a characteristic.

Non-destructive tests: test where the measured characteristic is not altered due to testing. Since the sample is not altered, multiple readings can be taken on the sample with the expectation of getting the same measured result. Destructive tests: test where the measured characteristic is changed due to testing. Since the sample is changed, there is no expectation of getting the same measured result over multiple readings.

So, in summary that makes up four types of MSA studies:

- Variable / Non-Destructive
- Variable / Destructive
- Attribute / Non-Destructive
- Attribute / Destructive Table

The following sections describe the requirements, measurement capability indexes and the typical acceptance criteria per MSA type.

General MSA Requirements

Test Environment Conditions - the test environment (i.e. temperature, humidity) should represent the conditions going forward. The effect of multiple environmental conditions can be evaluated if the study is properly designed and planned.

Sample Range - samples should cover the expected range of measurements.

Standard (for attribute MSA) - define the true answer (pass or fail). The standard is based on the inspection ratings of an expert opinion or a measurement system with known better inspection capability than the one under evaluation.

Measurement Instructions/ Training - follow the inspection instructions as defined in the controlled documents or redlines included with the protocol. Do not minimise variability by adding special instructions not defined in the controlled documents or redlines included with the protocol. Reference the controlled documents in the protocol. Special instructions are allowed when using pseudo samples provided that the variability is not minimised due to the instructions. Testers should have a high degree of skill and experience. Do not use new personnel or inexperienced people to conduct measurement studies.

Equipment Qualification and Calibration – The equipment must be calibrated prior to conducting the study. Evidence of the calibrated state should be documented in the report (e.g. calibration certificates etc.). It is important not to re-calibrate the equipment during the study as results can be different due to the calibration effect. The effect of calibration can only be evaluated if the study is properly designed.

Randomisation -
1. Assign the samples to the first operator in random order. Operator measures the parts.
2. Assign the samples to the second operator in random order. Operator measures the parts.
3. Assign the samples to the third operator in random order. Operator measures the parts.
Repeat the process described in steps 1 to 3 with the operators for a second and third trial.

Data collection - when documenting the results of a trial, the operator should not have access to the results from the previous trials. A different data collection sheet must be provided for each operator involved in each trial. In lieu of a different data sheet, a data recorder may be used to blind the data recording operator to the test data of previous runs.

Variable MSA Studies

Non Destructive/Variable MSA Studies

The key requirements for non-destructive and variable MSA studies include:
No. of Operators – at a minimum, 3 operators should be used during the study. More operators are also recommended if human/operator interaction is a source of measurement error.
Sample Size – a minimum of 10 units is recommended.
Trials - a minimum of 3 trials should be completed.

Destructive/Variable MSA Studies

If a test is destructive in nature, repeated measurements cannot be taken as the sample is damaged or destroyed as part of the test. One solution is to adopt standardisation of units where homogeneous samples are created by standardising the material or manufacturing process.

No. of Operators – 3
Sample Size – 10 units
Trials – 3 trials

This equates to 90 measurements in total. If standardisation is not feasible, the use of non-destructive pseudo-samples can be used. However, equivalence should be demonstrated between the pseudo sample and "true" units.

Attribute MSA Studies

Non Destructive

The recommended and minimum sample size requirements for attribute/non-destructive MSA studies are shown below:

Recommended Minimum Sample Size Requirement:
Operators - 3
Sample size - 25
Trials - 3

Destructive

When the test is destructive, repeated measurements cannot be taken as the sample is destroyed or altered. Some approaches are outlined below in order to quantify the measurement variability for destructive tests.

Standardisation Approach: homogeneous and representative samples are created by standardising the method of sample preparation, or material.

Sub-samples: cut each sample into three sub-samples to represent the three trials.

Pseudo-samples: create non-destructive pseudo-samples, documenting a rationale justifying the equivalence of the pseudo samples to the true samples.

Measurement Capability Index

The Measurement Capability Index (MCI) is calculated to assess the capability of the measurement system. The MCI is calculated as a % tolerance.

Measurement Capability Index acceptance criteria:
This index is used to evaluate the capability of the gauge to classify product against the specifications.

The index represents the % of the tolerance (upper specification limit (USL) and the lower specification limit (LSL) that is consumed by the measurement system variation.

Action Plan - Identifying Sources of Variation

Sources	Action
Measurement variation due to repeatability	Ensure the sample is not deformed over time due to repeated measurements and trails
	Ensure the equipment specification has the required precision
Measurement variation due to reproducibility	Review training and introduce standard work instructions

Suitability for Use

If an existing test method can be used with no or minor changes, a Test Method SFU Report can be used to document the test method validation.
A suitability for use report is appropriate only if the new product test method parameters are within the existing validated range.
If the test method parameters for the new product are outside of the validated range, the test method must be re-validated. Examples of cases which can utilise such suitability for use reports include:

- Test method transfer to a new manufacturing site.
- New product where the product specifications fall within the validated output range.
- Minor changes in component material which do not impact the validated test. Examples of changes that require full validation include:
- New products.
- Extension of product sizes that fall outside the validated range.

Test Method Validation Plan -Example

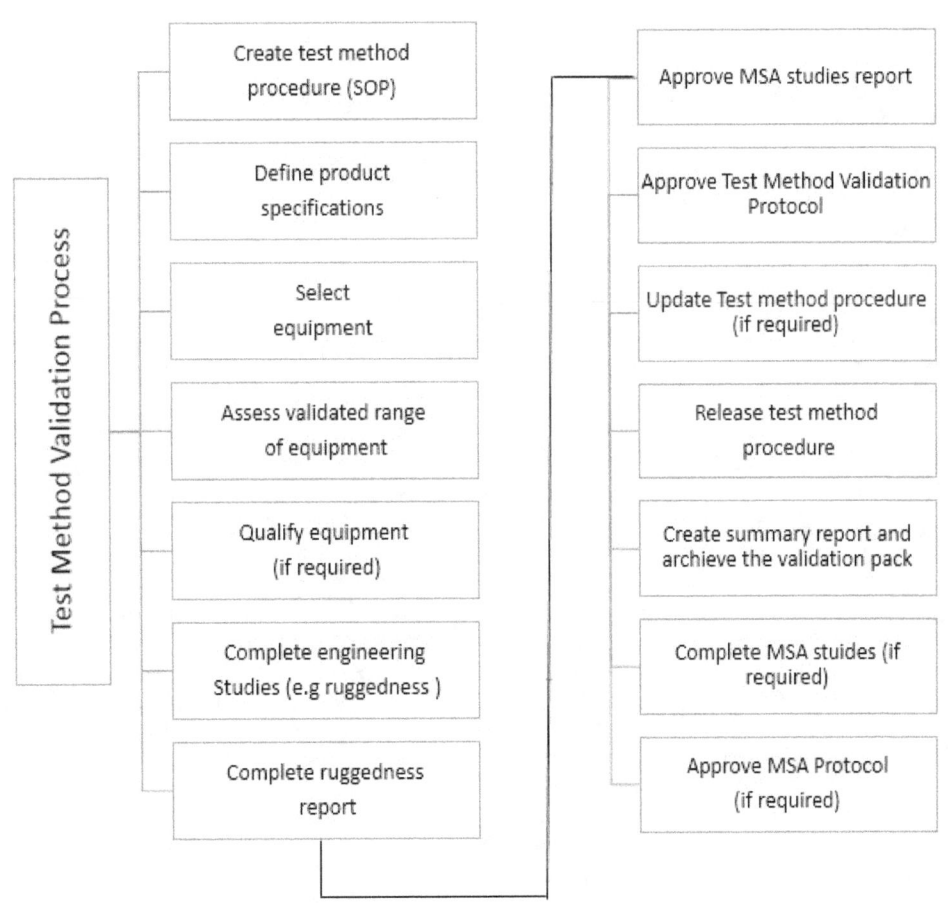

Figure 17: Example of TMV process showing key deliverables

CHAPTER 11

Supplier Validation

Introduction

Most manufacturers will require materials to be supplied to some degree, it may be in the form of stock materials that are processed or machined. It may be raw materials that are mixed and processed like in the food industry or pharmaceutical industry, or alternatively it may be bespoke components. Suppliers form a key part of the product lifecycle from the design stage where materials and their suitability are assessed through to the commercial manufacture and sale finished products. Organisations also depend on equipment manufacturers and suppliers for off-the-shelf systems and in many cases customised equipment and machinery. These scenarios also require suitable supplier management. Yet another area of supplier focus is storage containers, intermediate products, processing and manufacturing agents that are used.

Common requirements for all of the above components and their suppliers are the proper management, approval, supply and use of them in order to ensure that quality products reach the customer. Suppliers are evaluated to determine the adequacy and conformance of their quality system to applicable regulatory standards and internal company requirements. The breadth of any evaluation is dependent upon the service or material provided and the potential impact on the product quality or quality systems. Examples of items requiring supplier evaluation & approval include:

- ➢ Raw materials
- ➢ Components
- ➢ Packaging containers
- ➢ Intermediates
- ➢ Drums
- ➢ Chemicals
- ➢ Storage containers
- ➢ Manufacturing agents

The Goal of Supplier Evaluation & Approval

The main goal of supplier validation is to ensure only approved suppliers, capable of consistently supplying materials, products and services meeting defined standards are used. Before a supplier is used, the supplier and the manufacturing source, if different, should satisfy an assessment that demonstrates suitability. In addition to new materials, similar materials that are produced at a different site or using a different process should be assessed.

Why Does Supplier Quality Matter?

The quality, performance and safety of any product can be impacted by the quality of the components, raw materials and services applied in order to deliver a finished product. Poor quality may cause injury or illness to a consumer and be costly for a business. It can lead to (1) loss in earnings and customer dissatisfaction (2) legal action (3) loss of licensees (e.g. medical device manufacturers) (4) costly product recalls and so on. Taking the time to selecting the right supplier can mitigate against the risk of issues during the manufacturing process and commercial supply.

Medical Devices: ISO 13485, clause 4.1, requires manufacturers to identify the outsourced processes that an organization uses. It also requires that they identify the controls applied to those processes. The standard requires an organization to identify such processes in its quality management system (QMS) and to specify the application of the processes throughout the organization

Pharmaceuticals: US 21 CFR Part 211 requires purchasing controls to be applied to suppliers and 3rd parties. The intent of purchasing controls is to assist manufacturers in selecting only those suppliers, contractors, and consultants who have the capability to provide quality product and services

General Engineering: Many engineering companies will operate a quality management system such as ISO 9001.

Non Engineering: For business reasons, a supplier evaluation and approval process may be required.

Supplier Evaluation & Approval Process

Supplier Evaluation & approval is typically a multistep process. The complexity of the process can depend on several factors. Is the product or material supplied simple or complex? Take the example of simple material supply. A supplier may provide mild steel barstock of 15mm in diameter to a manufacturing firm. The receiving customer needs to consider the following as a minimum. Keeping in mind this is a relatively simple supply item.

Supplier Setup: this may simply be an administration task, where method of payment, terms or payment, delivery arrangements etc. are documented and agreed upon. These setup steps may be decided and planned but should not be introduced or activated until all the requirements have been met. E.g. quality requirements etc.

Product Quality: The customer may wish to receive a certificate of conformance or specification from the supplier with regard to the quality, composition and makeup of the material. Metals come in a variety of grades with different applications in mind. In some instances, customers may wish to receive materials from a particular manufacturer or country of origin.

<u>Delivery requirements:</u> The customer may wish to specify the manner in which the material is packed and shipped. Consignments that are too small or too large may create avoidable issues.

If we now look at a more complex supply requirement of a manufacturing company. A pharmaceutical company manufacturers an sterile injectable solution. Syringes are prefilled with the solution, packed and delivered to the customer. The delivery device, which is the syringe is sourced from an external manufacturer which supplies the assembly for filling. As this is a pharmaceutical product, patient and user safety are extremely important, therefore, the supplier evaluation and approval process are much more complex. Company requirements: The manufacturer should be certified to a Quality Management system e.g. ISO 13485.

Product Quality: As the syringe will store and deliver the solution dose, the quality of the materials, components and overall assembly of the syringe are critical to safety and functionality. The receiving customer will likely wish to see:

> Evidence of a Design History File (DHF) in relation to the syringe and delivery device
> Reports on material suitability with the proposed solution to be injected
> Testing and reporting on the biocompatibility, toxicology and bioburden levels of the product and packaging.
> Aging and stability studies to demonstrate the safe use of syringes are been stored.
> Test method validations for all physical, chemical and analytical testing
> Equipment validation for all equipment that is used in the manufacture of components
> Process Validation for process relating to manufacture and assembly

<u>Other requirements:</u> as part of the supplier approval process, an onsite audit may be required. At a minimum, a vendor assessment may be required as a desktop activity.

General Approaches to Evaluation

Faced with the evaluation and approval of a new supplier often involves a multi-disciplinary team. This can be made up of quality, technical, operations and EHS expertise. The first question to ask is what category the item to be supplied falls into. Some typical categories used to classify and evaluate items includes (1) Raw material, (2) Component / intermediate (3) Packaging Material and (4) Services. Organisations may also have their own internal categories that best suit their business operations.

However, each category should lay out a process that provides a structured list and sequence of events that are required to be completed in order to complete an appropriate evaluation and approval. If materials only relate to general cleaning, a throughout evaluation may not be required. However, if materials come into contact with products or can affect the quality of products, then the evaluation is a lot more in depth.

The level of evaluation can be divided into three distinct areas.

(1)Paper based assessments,

(2) Performance based assessments,

(3) Onsite audits assessments.

(1)Paper based assessments: this is the least rigorous form of assessment. It examines the supplier documentation such as specifications, CofAs, MSDSs and so on. While it is not as through as performance based assessments or onsite audits, it may require elements of the same to attain the level of confidence in the supplier. Often poor documentation provided by a supplier can be an indicated of more serious gaps of issues within a company. Therefore, a lot can be learned about a supplier by their response to documentation requests in support of any approval process.

(2) Performance based assessments: typically take more time to execute, performance based assessments are required for raw materials etc. that can impact product quality or safety.

(3) Onsite audit assessments: gives the customer a clear picture of a suppliers operations and internal processes. Onsite audits are usually completed in combination with performance evaluations and documentation reviews.

Guidance On Identifying Potential Suppliers

Before the purchasing process even begins, there are some simple steps that can be taken to shortlist suppliers and ensure they broadly fit the requirements of you as the customer. Firstly, a proper understanding of the need or requirement should be documented. If this is a simple consumable product then this may be a simple exercise e.g. supply of cleaning materials. However, if the supply of a bespoke engineering component is needed, that is manufactured offsite and inspected by the manufacturer, then a more detailed requirement is appropriate. (engineering drawings, functional requirements, material properties, specifications etc.)

You may also need to consult with your own internal technical staff to capture all of the right information. This prevents the embarrassing scenario of scoping a inadequate product or worse again, receiving a product or a service that does not meet the intended use. To identify the intended use of the product or service, consider the following points:

> What is the specific product or service you are buying?
> What impact does this have on your own business?
> What are the risks to your business and end user if you experience problems with this product or service?
> How can you be sure that the product or service you receive will actually meet your requirements?

The reputation and historical performance of suppliers may also be considered before any formal evaluation is initiated. If the supplier is certified to a Quality management system such as ISO 9001 or ISO 13485 or other, this may provide an extra level of confidence to doing business with a prospective supplier.

Definitions

Internal supplier: some companies or organisations may be able to source materials from within a company network. Also known as intra-company supply, there may be specific definitions that clarify the scope and applicability of internal suppliers at a corporate level. Also, regulatory bodies and accreditation bodies may have specific definitions and requirements. For example, 21 CFR Part 820 for medical devices, an internal supplier is when the supplier is under the same Quality System internal quality audit.

External supplier: has not connection with the receiving site and is not under the same Quality System internal quality audit.

Product: Product means components, manufacturing materials, in-process devices, finished devices, and returned devices. (21 CFR Part 820)

Approved Supplier/ Approved Status: A supplier who meets the requirements for Approved status as detailed below and who, therefore, can be used to supply a given material.

Approved Supplier List (ASL): a controlled list of suppliers that have been approved for use after an evaluation process.

Change Notification Agreement (CNA): A signed declaration that states that the Supplier agrees to notify the customer of any changes in its product or process. This allows the customer to determine whether the change may impact the quality of products.

Component: means any raw material, substance, piece, part, software, firmware, labelling, or assembly which is intended to be included as part of the finished, packaged, and labelled device. (21 CFR Part 820)

Compendial Organization: An organization certifying material standards that meet one or more compendial requirements (e.g. USP).

Conditionally Approved Supplier: A supplier may be conditionally approved and not fully approved as a result of an audit which results in critical findings or a number of major findings that need to be corrected. Continued use of suppliers with this status may include a requirement for additional controls and additional testing. It should also be recognized that this short of use comes with a degree of risk.

National Standards Organisations: nationally recognized organisations that create standards (e.g. NIST)

Potential Supplier: A supplier, identified for possible supply.

Conditional Status: An interim approval status given to a supplier during the supplier's evaluation period.

Restricted Status: where there is a potential risk to due to product or service related quality issues cor concerns. This may be a result of audit findings or multiple failures in business processes.

Supplier File: The summary file or collection of documentation that is generated detailing all information associated with a raw material, component, intermediate or packaging material.

Approved Status: A supplier status that meets the internal requirements of a company along with any regulatory or statutory requirements (e.g. relevant for medical components – must comply with ISO, regulatory requirements and competent authorities as required.)

Evaluation Types

Most supplier evaluations can be categorized into the following (1) Raw material, (2) Component, (3) Packaging Material and (4) Service.

Raw material: a raw material may come in granular from (chemicals, resins), sheet form (sheet metal, rolls of plastic) and so on. Typically raw materials are either processed or added to other materials in order to produce an intermediate product or final product. Raw material evaluation and approval involves some unique methods in order to ensure suitability for the intended use. The methods of materials science are commonly used to determine particle size, particle shape and density. Other tests may examine purity, water content etc.

Component: typically a part that is assembled along with other components to form a product. Components may be off-the-shelf items such as nuts and bolts, or it may be a part that is uniquely machined or manufactured for a specific use with a specific product and customer in mind.

Packaging Material: packaging materials can be broadly divided into primary and secondary packaging. Primary materials are the materials that house the product directly, and come into contact with the product itself. Examples include blisters packs of tablets. Secondary packaging refers to cartooning in which the primary packaging is placed. Supplier approval of packaging materials can be complex as not only functional and physical requirements need to be fulfilled, but regulatory and legal requirements may need to be incorporated also.

Service: services can range from cleaning and catering services provided to an organization, to consultancy and laboratory testing. Depending on the type of service, it can potentially impact product quality.

Supply Requirements for Raw Material

Overview

The below section provides a step by step requirements list for supplier evaluation and approval of raw materials from an external supplier.

Technical Specification Sheet (TSS): a TSS for raw materials provides detail on the technical, physical and chemical properties of a raw material, with technical information on the applications, restrictions, manufacturing methods etc. The prospective supplier should issue technical specification sheets in a controlled manner, with revision control and approval.

Material Safety Data Sheet (MSDS): an MSDS is a summary document that provides information on the composition, properties, safe instructions for use and potential hazards associated with a particular material. It lists the appropriate hazard and warning symbols that apply to the material. It is useful in understanding the risks associated with a raw material and helps to ensure appropriate precautions are implemented during the transport, storage and use of the material.

Certificate of Analysis /Certificate of Conformance: A CofA is a document that is certified issued by quality assurance that supplied goods meets its product specification. They commonly contain the actual results obtained from testing performed as part of quality control for an individual batch of a product.
Certificates should include:
(1) The name of the raw material.
(2) The batch/lot number.
(3) The grade of the raw material.
(4) Results for testing completed and acceptance criteria (limits). A statement whether or not the material was found to comply with the relevant specifications.
(5) Company name, full address and contact details of the site of manufacture.
(6) The date of manufacture (DOM), expiry date and recommended retest interval if applicable.
(7) Finally, the appropriate quality authorisation for release of the batch.

Samples: As part of any raw material evaluation and approval, obtaining samples is standard and common sense. The term "sample" has multiple meaning. For example if you have raw materials that are used in the finished product e.g. sugar as part of a soft drink, then raw material samples should be obtained and tested to ensure it meets the required specifications. Furthermore, samples may be required in order to manufacture test batches to ensure the raw material is compatible within the process and the quality of the finished product is not impacted. When obtaining samples, it is best practice to receive a minimum of 3 distinct lots or batches.

Test Methods: Physical, chemical and Analytical testing methods used by the supplier to test the raw material.

Labelling Requirements: The labelling details should be communicated to the supplier or alternatively the standard set by the supplier should be reviewed and approved.

Delivery Requirements: Delivery requirements should be agreed with the supplier as part of the evaluation process. Raw materials may be shipped in bulk containers that require special handling equipment.

<u>Supplier Packaging and Storage Instructions:</u> This step ensures that materials are supplied in appropriate containers that are suitable for use. The supplier should provide document information on how the material is packaged, dimensions, type of packaging material etc. For raw materials, this is an important factor, as it prevents contamination. Statements of suitability for use may be required if pharmaceutical of food stuffs are stored. The potential supplier should detail any special requirements for storage or any maximum storage interval for the materials. Finally, the supplier should detail how the material is packaged, overall dimensions and the type of packaging material.

<u>Toxicology assessment (as required):</u> A toxicology report or statement may be required for certain products to ensure the material is assessed for compliance to market and regulatory requirements.

Supply Requirements for Components

This section details the initial requirements relating to the supply of a component manufactured externally.

Process Technologies: The supplier/manufacturer should provide details on the equipment and processes employed to manufacture the component.

Technical (Engineering) Drawing: an engineering drawing showing all dimensions and tolerances of the part. Details on other critical to quality attributes such as surface finishes etc. should also be included. The engineering document should be revision controlled

Technical Specification Sheet (TSS): a TSS may be used to provide a summary of component details and applications.

Material of Construction certificates (MOC): materials of construction provide evidence that materials meet the required specification and grade. For example, stainless steel comes is various grades 316L, 304, 305. This gives the customer confidence that the right materials have been supplied and provides documentary evidence if required as part of the product or service.

Material Safety Data Sheet (MSDS): while an MSDS is more applicable to raw materials and manufacturing agents (e.g. IPA, oils and lubricants), an MSDS for a primary material of a component.

Certificate of Conformance: A CofC for a component ensure that the dimensions and features of the component confirm to engineering specifications and requirements.

Certificates should include:
(1) The name of the component
(2) The batch/lot number.
(3) Results for testing completed and acceptance criteria (limits). A statement whether or not the samples were found to comply with the relevant specifications.
(4) Company name, full address and contact details of the site of manufacture.
(5) The date of manufacture (DOM), expiry date and recommended retest interval if applicable.
(6) Finally, the appropriate quality authorisation for release of the batch- quality approval

Test Methods: Physical, chemical and Analytical testing methods used by the supplier to test the component should be available to the receiving customer.

Supplier Packaging and Storage Instructions: All components should be supplied in appropriate containers that are suitable for use. The supplier should detail how the parts/components are packed, e.g. loose in a box, loose and bagged, loose and double bagged.

Labelling Requirements: While components may not be individually labelled, the drums of packaging needs to identify the product, product code, batch no and other information that provides traceability and identification.

Delivery Requirements: Delivery requirements should be agreed with the supplier as part of the evaluation process. Components may be shipped in bulk. The total quantity or total number of batches may be determined by the size of components and the restrictions of pallets.

Supply Requirements for Packaging

Overview

Packaging materials such as blisters, polybags, cartons have a barrier function for products. Therefore the intended use and type of product will direct the level of evaluation required.

Technical Specification Sheet (TSS): if the packaging material is a standard industry use material, it is likely the manufacturer/supplier will have a TSS for prospective customers.
Material Safety Data Sheet (MSDS): for medical devices, food and pharmaceutical products, an MSDS is a basic requirement that forms part of the Design history file.

Certificate of Conformance: packaging processes such as blister packing and blister sealing are sensitive processes. Therefore, all materials used in creating the packaging system need to meet a predefined specification e.g. thickness, tensile strength, melting point etc.
Samples: Packaging systems are often made up of multiple elements or materials. For example medicinal tablets are often blister packed in pvc and foil. Therefore, when sampling a potential supplier different batches of each element should be matched with one another. For example 3 batches of foil material and 3 batches of pvc material would challenge the variation between batches.

Test Methods: Tensile testing and thickness of packaging materials are important factors in ensuring the integrity of packaging systems. Many tests can be conducted in accordance with international standards. E.g. ASTM, ISO.

Toxicology assessment (as required): A toxicology report or statement may be required to ensure the suitability of packaging materials.

The following diagram provides an overview of a simple supplier evaluation process.

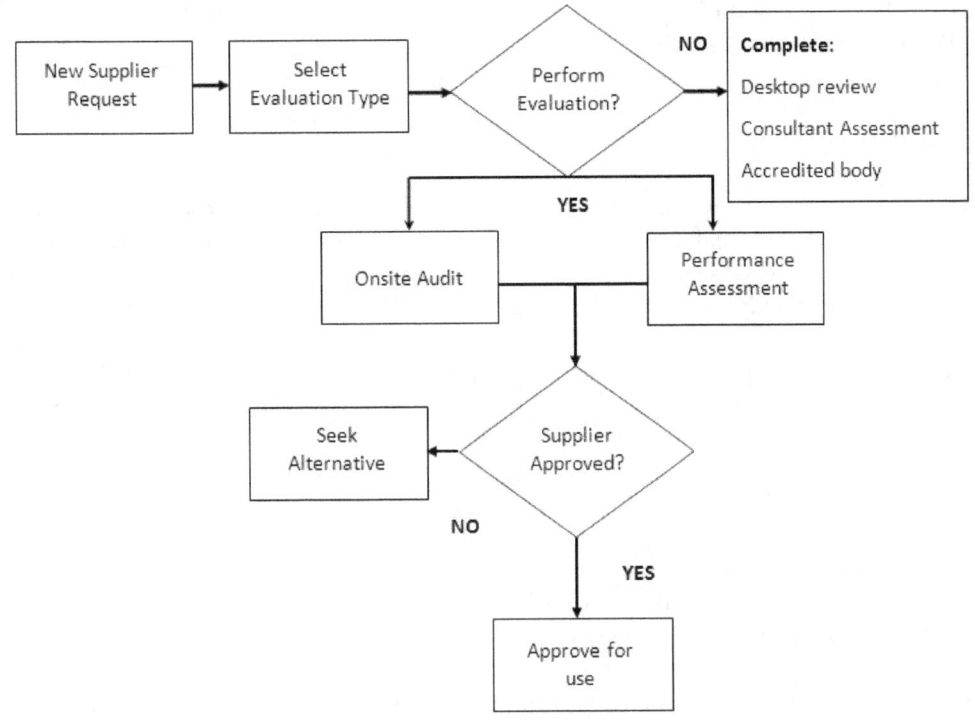

Figure 18: Sample Supplier Evaluation Process

After a new supplier request is identified or made, it should follow a set process of evaluation and approval. Typically, organisations will have a procedure to govern the process. Having a process flow is also a useful aid for those tasked with applying a supplier evaluation and approval process. In the above example, the second step in the process is to select an evaluation type. Individual organisations may have their own categories and tools in order to choose the right category for the supply issue at hand. Generally, the following categories facilitate most supply scenarios-(1) Raw materials, (2) Components, (3) Packaging Materials, (4) intermediates and (5) Services.

The evaluation type and the criticality of the supply item will determine whether or not a performance evaluation is required. The decision to complete a performance evaluation is often documented to provide a transparent decision history. For example, if a component is externally supplied, the supplier must demonstrate that the manufacturing process are controlled and validated. The performance of the component in the finished product may also need to be tested and documented. For physical products, a risk assessment is useful to determine the criticality of components and what level or evaluation of the supplier is required. Risk assessments can include Design Failure Mode Effects & Analysis (DFMEA), Process Failure Mode Effects & Analysis and Fault Tree Analysis.

Onsite audits provide an opportunity to review many aspects of a suppliers business. From the quality of the building, internal conditions, housekeeping, layout, equipment, work practices, documentation, and storage conditions.

Alternative Suppliers

An obvious reason for the need for alternative suppliers is balance the risk associated when a sole supplier is used. Sole suppliers represent an inherent risk to an organisations operations and commercial activities. If that sole supplier goes bust or experience internal supply issues this can lead to continuity of supply problems for their customers. Other issues may arise such as increases in price or changes in the terms of delivery. Even the development or reoccurrence of quality issues can call in to question the use of a supplier and may prohibit their use. Therefore, evaluating and approving alternative suppliers has value and mitigates the risks described above.

Sample Checklist for Supplier Evaluation and Approval

The below is a generic checklist that can be applied for new suppliers of materials/components or other categories. It can be adapted for the task at hand. Some requirements may not be required depending on the product or service.

Equipment: Is the Equipment Qualified (Installation & Operational Qualification- Equipment), is the equipment fit for the intended use? (IQ/OQ ensures the equipment is installed correctly and operates as intended. Critical functions such as equipment based instruments will require calibration)

Process: has the process been implemented and proven it can consistency deliver the outputs required? For medical devices, an Operational qualification should be performed. This ensure the process can deliver quality product that meets all specifications at the extremes of the process (the upper and lower process settings).
Performance Qualification examines the manufacture or processing of products under normal anticipated conditions e.g. the manufacturing conditions that are expected when the supply of items comes online.

Measurement & Test methods: if the supplier provides raw materials or a component, what measurement systems are used, are the systems qualified and fit for purpose? Are test methods suitable, performed to an international standard.

Documentation: Ensure the supplier file is complete and accurate. Refer to the previous section on Supplier file content for guidance.

Risk assessments: if an alternative supplier is required, new risk assessments or updates to existing risk assessments may be required. After any risk assessments have been conducted, ensure all mitigations and control actions are completed.
The may also have completed risk assessments in order to demonstrate that their processes are in control.

Audits: if applicable and if highlighted as a requirement, ensure any findings or recommendations as part of any audits are appropriately closed and verified.

RTM: documented the requirements (URS) of the system and are the requirements traceable from a source (URS) to the qualification or validation activity (IQe, OQe, OQ p, PQ p)?

The Supplier File

A Supplier File is a formal way of organizing and archiving information and documents relating to a specific supplier and a specific item. The supply item may be a raw material, a component, a packaging material or a service.
A Supplier File may include the following:

> - Technical specification sheet / data sheet.
> - Certificate of Suitability /Statement of suitability
> - Supplier Questionnaire
> - MSDS
> - Quality Agreement
> - Technical Specification Sheet
> - Sample Cert of Analysis's or sample Cert of Conformance
> - Testing methods
> - Supplier packaging information and storage conditions
> - Labelling specification
> - Delivery specification and methods
> - Quality Agreement

Advantages:

Ensures a common depository is maintained for the material/item and supplier
Ensures retrieval of information during audits is timely and organized
Facilitates the up keep of documents and data as a result of changes

Disadvantages:

Requires a gatekeeper to manage the file
The system is only effective if it is maintained properly, this can take time and resources

A fully signed-off complete supplier file containing all the information required for "Approved" status should completed as soon as possible in a project.

Changes to Supplier Files

In the lifecycle of a product or service, is there naturally changes that can occur of time. Most of these changes will be internal within an organisation and may not affect suppliers or vendors. However, in some cases, changes may come from the external supplier. This can be a result of changes to their manufacturing processes, changes to their suppliers, sale of assets, mergers or acquisitions.

A change in a raw material or component specification needs to be assessed and evaluated to ensure that there is no adverse impact on the product or service. Likewise any change that is implemented or followed through, must incorporate the new data or specifications in the supplier file.

Considerations for Delivery & Labelling

Window of delivery

All deliveries should arrive on the day specified as agreed on delivery schedules. Early or late deliveries cause confusion and not having the correct resources available may result in a delay to receipt of delivery.

Delivery notes

Typically delivery notes are included with documentation when a shipment is been prepared. The content of delivery notes should be agreed upfront with the supplier. The following can be considered for inclusion in a delivery note.

- Material name
- Item number of item code
- Purchase Order # and
- Supplier lot #
- Supplier description of the product
- Total quantity
- Number of pallets in the delivery
- Quantity of items per pallet
- Type of pallet used in delivery Euro or Standard pallet)

Pallets & Labelling

Considerations for delivery pallets and labelling include:
- The size of labels displayed e.g. A5 size or A4 size
- What information is displayed on the label:
- Item description, Item number or code
- P.O. Number,
- Supplier Name

- Quantity,
- Supplier Lot Number

The type of pallet should be agreed before the commencement of delivers. E.g. Euro pallets (800x1200mm) or Standard pallets (1200mm x 1000mm). A minimum standard of cleanliness of pallets should also be provided. (No damage, no missing struts, no oil markings etc.). The supplier should also detail the maximum weight and height of pallets and ensure no overhang on pallets.

Documentation

The following are the delivery documentation required with each delivery
- Delivery invoice
- Certificate of Analysis or Certificate of Conformance with the goods

CHAPTER 12

Summary of Good Manufacturing Practices

Introduction

Good Manufacturing Practices are a set of practices that are required in order to comply with industry standards and regulations. GMP helps to minimise the risks involved during manufacturing and helps to ensure products meet quality and regulatory standards. A GMP quality system ensures that products are consistently produced and controlled according to predefined quality standards. It is designed to minimise the risks involved in any pharmaceutical production that cannot be eliminated through testing the final product.

Often, a broader term is used in industry -GxP-where the "x" is used as an umbrella letter representing different subjects or disciplines in industry. Some prime examples include GLP (Good Laboratory Practice), GDP (Good Documentation Practice), GEP (Good Engineering Practice) and GMP (Good Manufacturing Practices). Furthermore, the use of a lower case "c" as a prefix indicates "current" or "up-to-date". So cGMP stands for "Current Good Manufacturing Practices.

This means that some conventions or practices are subject to change within the industry. Therefore, it is important to be up-to-date in the application of cGxP or cGMP

There are multiple regulators and organisations that provide definitions of "Good Manufacturing Practices". They include Organisations such as the World Health Organisation (WHO) and the International Society of Pharmaceutical Engineering (ISPE). Other definitions are offered by bodies such as the American competent authority for Food and Drug Administration. It is good to have an awareness of how organisations, bodies and competent authorities define GMP, and one should always review the "local" regulatory landscape. Below some definitions are provided to provide a feel for GMP and highlight the common thread between definitions.

W.H.O. World Health Organisation-"Good Manufacturing Practices (GMP, also referred to as 'cGMP' or 'current Good Manufacturing Practice') is the aspect of quality assurance that ensures that medicinal products are consistently produced and controlled to the quality standards appropriate to their intended use and as required by the product specification."

Food and Drug Administration: cGMP refers to the Current Good Manufacturing Practice regulations enforced by the US Food and Drug Administration (FDA). cGMPs ensure systems are properly designed and monitored, safeguarding the control of manufacturing processes and facilities. Adherence to the cGMP regulations ensures the identity, strength, quality, and purity of drug products by requiring that manufacturers of medications adequately control manufacturing operations. This includes establishing strong Quality Management Systems, obtaining appropriate quality raw materials, establishing robust operating procedures, detecting and investigating product quality deviations and maintaining reliable testing laboratories. This formal system of controls at a pharmaceutical company, if adequately put into practice, helps to prevent instances of contamination, mix-ups, deviations, failures and errors. This assures that drug products meet their quality standards.

MHRA (Medicines and Healthcare Products Regulatory Agency) defines GMP as follows:

"Good Manufacturing Practice (GMP) is that part of quality assurance which ensures that medicinal products are consistently produced and controlled to the quality standards appropriate to their intended use and as required by the marketing authorisation (MA) or product specification. GMP is concerned with both production and quality control. Many of the drivers of GMP in effect are also benefits to the manufacturer. Good manufacturing practices are an expected practice in regulated industries and a manufacturer must meet all relevant GMP regulations if they wish to manufacture within a country or sell to a particular market. It is important to maintain accurate, complete, up-to-date and consistent information to ensure patient safety and reduce any potential risks."

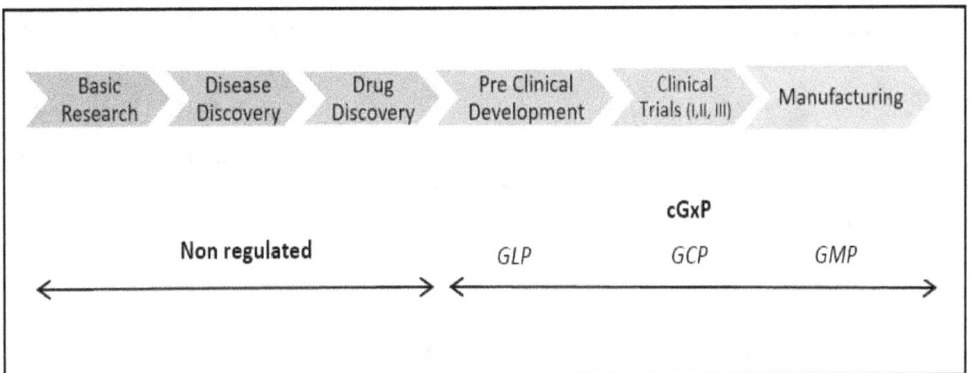

Figure 19: A simple Lifecycle overview showing where GxP and GMP is implemented.

History Of GMP

1906: The Pure Food and Drug Act: Establishment of regulatory agency- FDA. The act makes it illegal to sell "adulterated" or "misbranded" food and drugs.

1938: Federal Food, Drug and Cosmetic (FD&C) Act: Sulfanilamide containing poisonous solvent causes 107 deaths.

1941: Insulin Amendment: requires FDA to test and certify purity and potency of insulin.
1962: Kefauver-Harris Drug Amendments Tragedy: Thalidomide tragedy causes severe birth defects in thousands of European babies.

1978: CGMPs Final Rule for drugs and devices (21 CFR Parts 210–211 and 820) establishes minimum Current Good Manufacturing Practices for manufacturing, processing, packing, or holding drug products and medical devices.

1979: GLPs Final Rule (21 CFR Part 58) -Good Laboratory Practices (GLP) for conducting non-clinical laboratory studies that support applications for research or marketing permits for human and animal drugs, medical devices for human use and biological products.

1982: Tamper-resistant packaging regulations issued for over-the-counter medicinal products: Acetaminophen-capsule poisoning by cyanide causes several deaths.

1987: Guideline on General Principles of Process Validation are issued by the FDA.

1996: Proposed Revision to US CGMPs for Drugs and Biologics (21 CFR Parts 21–211) adds detail for validation, blend uniformity, prevention of cross-contamination and out-of-specification results.

1997: Electronic Records Final Rule (21 CFR Part 11) requires controls that ensure security and integrity of all electronic data. 1998 Draft Guidance on Manufacturing, Processing, or Holding Active Pharmaceutical Ingredients and Investigating Out-of-Specifications and Test Results.

Consequences Of GMP Failures:

FDA Warning Letters

This is an official message from the United States Food and Drug Administration (FDA) that usually states that it has found that a manufacturer or other organisation has violated some rule of the Quality System regulations.

FDA 483s

An FDA 483 letter typically includes a summary of findings and observations in relation to an audit or inspection where the FDA representatives have reason to believe GMP or other regulations have been violated or are not being met. In response to an FDA 483 letter, the company should address each item and provide a timeline for correction or request clarification of what changes are required.

Below are the top three items of concern:

- Procedures not written or are not fully followed
- Poor investigations of discrepancies or failures (CAPA)
- Absence of written procedures

Consent Decree

A consent decree is a binding order issued by a judge that stipulates the voluntary agreement by the participants in a case of litigation. Decrees are sometimes issued after one party voluntarily agrees to cease a particular action without admitting to any illegality of the action to date.

Product Recall

Product recalls can be initiated in circumstances where a manufacturer becomes aware of a manufacturing defect, packaging or mislabelling defect or otherwise. Product recalls can have a serious impact on the financial stability and future of a company due to bad publicity and loss of sales.

Plant Injunction

An injunction is a judicial process initiated to stop or prevent violation of the law, such as to halt the flow of violative products in interstate commerce and to correct the conditions that caused the violation to occur. (FDA 21 U.S.C. 332; Rule 65, Rules of Civil Procedure). If a firm has a history of violations and has promised correction in the past but has not made the corrections, the injunction is more likely to succeed. However, the freshness of the evidence is critical.

For an injunction action to be credible in the eyes of the Department of Justice (DOJ), the U.S. Attorney and the court, the evidence must be current. Timeliness is an important factor when considering an injunction action, with or without a Motion for Preliminary Injunction or a temporary restraining order (TRO). However, case quality and credibility must not be sacrificed to meet guideline time frames. The purpose of the guideline time frames is to limit, as much as can reasonably be expected, the need to update evidence. Updating entails extra work at all levels of the case development and review process and more importantly, delays obtaining an injunction which is intended to stop violations that adversely affect the safety or quality of products in commerce.

When an injunction is granted, FDA has a continuing duty to monitor the injunction and to advise the court if the defendants fail to obey the terms of the decree. (FDA 6-2-Injunctions)

Debarment

The FDA has the authority to "disqualify," or remove, researchers from conducting clinical testing of new drugs and devices when the agency determines that the researcher has repeatedly or deliberately not followed the rules intended to protect study subjects and ensure data integrity. Further, the FDA can disqualify a clinical investigator who has repeatedly or deliberately submitted false information to the agency or study sponsor in a required report.
Under its statutory debarment authority, the agency may also ban, or "debar" from the drug industry individuals and companies convicted of certain felonies or misdemeanours related to drug products. Once individuals have been subjected to "debarment," they may no longer work for anyone with an approved or pending drug product application at FDA. Debarred companies may no longer submit abbreviated drug applications.

What Is Quality?

Quality can be defined as the ability to consistently produce products meeting the same specifications time after time. Products must be safe, pure, uniform and effective. Specifications can be set down internally within a company, however, depending on the product, external specifications from regulators or standards may be required.

Patient safety is the primary focus of any pharmaceutical drug or medical device. This is the expectation of any patient or user. Secondly, the patient or user is interested in receiving an effective product. It is product specifications that ensure these criteria are accounted for.

What Is A Quality Management System?

A Quality Management System, often abbreviated to (QMS) is any system based on a collection of business processes that are primarily focused on providing safe and quality products that consistently meet customer requirements. The core themes of a QMS are outlined below.

Customer and Regulatory Focus

An understanding of the customer needs and requirements should be evident within the organisation and with the future vision of the company. The company should have an understanding of the regulatory landscape as this is subject to change over time. In turn the company should be positioned to respond to that change.

Leadership

To truly lead, one must be accepted in the hearts and minds of those they lead. Authentic leadership pays off. A leader should foster a sense of togetherness and common vision. A leader is anyone who influences or directs people either formally or informally. We are all leaders to some extent.

Involvement

Engagement by everyone across an organisation is now recognised as being key in the successful deployment of any Quality Management System. Everyone should have a voice within the company. As the saying goes "we are only as strong as the weakest link" is very apt.
The Process Approach

Systems Management

This essentially means that systems are defined and described in writing along with the appropriate responses to expected issues that arise. Effective systems management must ensure that the various systems work in support of each other and communicate effectively with one another.

Decision Making

In order to make the right decision, the person empowered to make the decision must be informed. To be correctly informed one must have the correct details and facts available. In a manufacturing environment the facts are essentially data and the analysis of data. During manufacturing or processing, data is generated as a result of monitoring and measurement of products and the related processes.

Supplier Management

Don't ruffle your suppliers' feathers. Security of supply is key in delivering products to customers or patients again and again, Raw materials or sub-components sourced from external suppliers must always be sourced at the right price and time with the emphasis on getting the best quality possible.

Continuous Improvement

For ISO 13485 continuous improvement refers to improving the effectiveness of the Quality Management System. It is harder to drive improvement of the product due to regulatory and practical requirements.

The key elements of a QMS are listed below.

Quality Policy: A company will document their commitment and approach to quality within their organisation. It usually sets out how they plan to achieve a high and consistent standard of quality. It should in some way speak to the customer or end user.

Quality Objectives: Quality objectives can be documented in a Quality Plan at site or organisational level. An effective way of defining quality objectives is use of the SMART method. SMART stands for Specific, Measurable, Achievable, Realistic and Timely.

Quality Manual: An in-house guidance document to provide a framework for achieving the quality objectives.

Organisational Structure and Responsibilities: Organisational charts can be used to map out the company structure. Roles and responsibilities can be documented in site quality plans, job descriptions and Standard Operating Procedures.

Data Management: A coherent approach to the provision, storage and maintenance of data.

Processes: Processes are defined and documented.

Resources: Resources must be properly understood, allocated and linked across the organisation.

Product Quality & Customer Satisfaction: The proper management and investigation of complaints is important to reduce future instances from reoccurring. Continual engagement with the end user or customer is critical.
Continuous improvement including corrective and preventive action- where continuous improvement projects and initiatives are encouraged and supported. The application of a CAPA system to ensure quality is maintained and consistent.

Maintenance: A Preventative Maintenance schedule is in place and managed accordingly.

Sustainability: All work practices are sustainable and consistent throughout the lifecycle of processes and products.

Auditing: Systems are auditable and maintained to allow internal or external review and audit.

Engineering Change Control: Where changes are required to validated processes or equipment, changes are managed and introduced under change control.

A common acronym used to highlight the aims of Good Manufacturing Practices (GMP) is SPUE which stands for Safe-Pure-Uniform-Effective. This definition is particularly suited to pharmaceutical products as the chemicals and drugs used need to be pure and free of contaminants. Furthermore, they need to be uniform, meaning they will have the same constituents from tablet to tablet and batch to batch. A description of each word is shown below:

SAFE- the product has the right ingredients if it is a drug product. It is packaged as intended and correctly labelled in order to provide identification and safe use.

PURE- it is free of contaminants, foreign matter, chemicals and harmful microbes.

UNIFORM- The product is manufactured consistently and will have the same quality between batches manufactured on different days.

EFFECTIVE- Ultimately, the product must be effective in treating the medical condition. To be effective, it requires the correct ingredients, the correct amount of ingredients and correct packaging to maintain the product stability over time.

Ten Rules for GMP

Rule #1: Get the facility design right from the start

Rule #2: Validate processes

Rule #3 : Write good procedures and follow them

Rule #4 : Identify who does what

Rule #5: Keep good records

Rule #6 : Train and develop staff

Rule #7 : Practice good hygiene

Rule #8 : Maintain facilities and equipment

Rule #9: Build quality into the whole product lifecycle

Rule #10: Perform regular audits

Rule #1: Get the facility design right from the start

Every food, drug, and medical device manufacturer aims to operate their business in accordance with the principles of Good Manufacturing Practice (GMP). It's much easier to be GMP compliant if the design and construction of the facilities and equipment are right from the start.

It's important to embody GMP principles and use GMP to drive every decision.

Facility layout:

Lay out the production area to suit the sequence of operations. The aim is to reduce the chances of cross contamination and to avoid mix-ups and errors. For example, don't have final product passing through or near areas that contain intermediate products or raw materials. Aim to:

- remove unnecessary traffic in the production area
- segregate materials and products
- minimise potential for mix-ups and errors

Facilities Design:

The following points should be considered at the facility design stage. The impact of choices and decisions here can must be understood. The scope and type of manufacturing will determine many of the building and facility requirements. Risk assessments should be considered as a tool in identifying the right materials and design features.

Materials of construction
Windows
HVAC requirements
Utilities to be supplied
Emergency generators/UPS systems
Access to site and area's
Entrances

FDA Requirements

211.42 Design and Construction Features
(a) "Any building or buildings used in the manufacture, processing, packing, or holding of a drug product shall be of suitable size, construction, and location to facilitate cleaning, maintenance, and proper operation."

211.46 Ventilation, Air Filtration, Air Heating and Cooling
(b) "Equipment for adequate control over air pressure, micro-organisms, dust, humidity, and temperature shall be provided when appropriate for the manufacture, processing, packing, or holding of a drug product."

Equipment:

Design, locate, and maintain equipment to suit its intended use. The equipment should be:

- easy to repair and maintain
- designed and installed in an area where it can be easily cleaned
- consistent with the intended use
- calibrated at defined intervals (if required)

Environment:

It's important to control the air, water, lighting, ventilation, temperature, and humidity within a plant so that it does not impact product quality. You should design facilities to reduce the risk of contamination from the environment.

Make sure that:

- lighting, temperature, humidity and ventilation are appropriate
- walls, floors and ceilings are smooth, free from cracks and do not shed particulate matter
- interior surfaces are easy to clean
- pipe work, light fittings, and ventilation points are easy to clean
- drains are sized adequately and have trapped gullies.

Rule #2: Validate processes

Validation is defined as "Establishing documented evidence that provides a high degree of assurance that a specific process will consistently produce a product meeting its pre-determined specifications and quality attributes." It's a GMP requirement to prove control of the critical aspects of certain operations. New facilities and equipment, as well as significant changes to existing systems, require validation.

All validation activities should be well planned and clearly defined. This is usually by means of a Validation Master Plan, or VMP. Before you get to this stage consider all the critical parameters that may be affected and impact product quality; what happens if the stirring speed is changed? How does this affect temperature or pH? Once this is complete, define the testing and documentation required. Validation usually is made up of three components:

Installation Qualification, or IQ, which is testing to verify that the equipment is installed correctly as per manufacturers recommendations.

Operational Qualification, or OQ, which is testing to verify that the equipment operates correctly as described in the user requirement specification.

Performance Qualification, or PQ, which is testing to verify and confirm that product(s) can be consistently be produced to specification under anticipated conditions.

The FDA defines 4 types of validation. (1) Prospective, (2) Concurrent, (3) Retrospective and (4) Revalidation. These various types of validation form what approach the validation takes. E.g. is it a new process that will be validated in advance of commercial manufacturing, or, will the process be validated in a staged basis – concurrently. Etc. The 4 types of process validation are explained in the definitions section.

Rule #3 : Write good procedures and follow them

Within a regulated company, many documents are used to instruct, track, test and record information on the manufacturing process. Any document that can impact the quality of the product or product safety is treated as a controlled document. A controlled document is classified as a legal document. These controlled documents must incorporate certain requirements such as the date of approval, revision control and appropriate levels of review and approval. The accuracy and content of these documents can be subject to review by regulatory bodies Including the FDA in the US and the MHRA in the UK It is important that there are no errors or "questions marks" over the content. Examples of controlled documents include:

- Policies
- SOPs
- Specifications
- MFR (Master Formula Record)
- BMR
- Validation protocols and reports
- Forms
- Logbooks
- Records
- Bills of Materials (BOMs)
- Test Methods

Written procedures are controlled documents that provide detailed step-by-step instructions for the user. Written procedures promote consistency as they allow the same task to be performed in the same way, even by different people. They also act as a reference. If changes or improvements are identified, having a procedure in place creates a clear starting point which can be improved or modified in a controlled manner.

-Procedures should be written using clear and concise language
-Steps should be numbered clearly and individually to make them easy to follow
-Remember, written procedures are only effective if they are followed correctly, consistently and at all times by everyone

Never deviate from written procedures, these controlled documents ensure consistency and accuracy is maintained over time.

Documenting Work:

As the saying goes "Good Science starts with good documentation", meaning that data should be recorded as soon as available and accurately, following principles of GDP (Good Documentation Practices). The rule of thumb "If it's not documented, it didn't happen" is an important statement. To provide evidence or proof that a process is producing quality product, it must be verifiable in a document. Documentation requires that you record, sign and date every step of the operation you perform. It is a regulatory requirement for companies manufacturing medical products to keep accurate documents and records relating to products and their manufacture. To ensure documents meet a high standard of compliance both prompt recording and accurate recording is paramount. Prompt recording- recording of test results and data should take place as soon as possible and as soon as available. This is to minimise risk or errors with late entries which could lead to quality issues or audit findings.

Rule #4 : Identify who does what

All employees understand what tasks and activities need to be done each day. Documenting these responsibilities is key to ensuring people fulfil their roles and responsibilities. Job descriptions for each role and employee should be created and should detail the following:

- job title
- job objective
- duties and responsibilities
- skill requirements

An organisational chart helps to document and display roles and functions.

Rule #5: Keep good records (Documentation)

Good science starts with good documentation. As part of GMP it is essential to keep accurate records, and during an audit, it helps convey that you are following procedures. It also demonstrates that processes are known and under control. Guidelines on Good Documentation and record keeping:

- Record all information immediately upon completion of a task

- Write legibly with indelible ink. By signing records you are certifying that the record is correct and that you have performed the task as per the defined procedure

- Correct mistakes as per GDP. Draw a single line through any error, and initial and date the correction. Include a reason for the correction at the bottom of the page if the reason is not obvious

- Record details if you deviate from a procedure. Ask your supervisor or the Quality Department for advice should a deviation occur

- If it's not documented then it didn't happen!

Rule # 6: Train staff

Training should be provided for all employees who work within manufacturing, production or laboratories or where they may have an impact the quality of the product. Basic training on GMP is a fundamental requirement, followed by any job specific training as required. All training should be documented with a record of when the training occurred, dates, attendees and results of any assessments. Training records are often drawn upon by external auditors and are an essential record.

Rule #7 : Practice good hygiene

In order to reduce the risk of product contamination, good hygiene by every person is required. A culture of hygiene awareness should be evident in a GMP compliant facility. The practice of good hygiene should be supported by procedures and monitoring programs.
Depending on the classification of medical device been manufactured, the level of cleanliness required increases as the risk to patient increases. Therefore, sterile product manufacturing will require more strict controls and levels of hygiene.
Points to note:
Practice good personal hygiene by washing your hands at regular intervals
Wear the required PPE and protective garments and follow gowning procedures
If you are ill, inform your manager
Minimise any contact with product or product contact surfaces and manufacturing equipment.
Do not eat or drink in GMP areas or where indicated

Rule #8 : Maintain facilities and equipment

Preventative maintenance plays an important role in ensuring facilities and equipment remain fit for purpose. Regular maintenance prevents equipment breakdowns and unplanned interventions which disrupt production and cause backlogs within the process. Proper maintenance also reduces the risk of product contamination and helps to maintain the 'validated state' of the equipment and facility. GMP requires accurate records relating to maintenance activities are kept for audit and quality purposes.

Rule #9: Build quality into the whole product lifecycle

Quality must be paramount in any medical device. From its very design through development and into commercial manufacturing, quality must meet acceptable levels. Quality is the responsibility of everyone, not just the quality department.

Control of Components:

All materials and components when accepted onsite must meet predefined acceptance criteria. Sampling and accepting testing may be required. Typically, suppliers provide certificates of conformance to ensure the materials or components meet specifications. Suitable storage conditions need also be accounted for. All materials and components must be approved prior to release for manufacturing. Rejected materials or materials that fail inspection should be identified and stored in a secure area to prevent unauthorized use.

Control of the Manufacturing Process:

Establish records and procedures to ensure that employees perform the same job every time. Each product must have:
- A master record that outlines the specifications and manufacturing procedures.
- Individual batch records to document conformance to the master record.
- SOPs and procedures for cleaning and maintaining the equipment and areas.

Packaging and Labelling Controls:

Proper Packaging and labelling is necessary to identify how materials are packed, stored and labelled. Distinctive labels and accurate descriptions help prevent mix-ups and errors. Labelling also supports traceability, to different batches or different products.

Rule #10: Perform regular audits

Audits are conducted to assess if GMP rules and regulations are followed. External bodies such as the Food and Drug Administration (FDA) conduct formal audits also known as external audits. A Corrective Action Preventative Action (CAPA) system is required to manage and fix issues identified during an audit.
Simply put, an audit is a review activity that examines if a company or organizations processes are been followed. It also allows the identification and improvement of any concerns or non-compliances. Audits can be internal (conducted by internal staff) or external, (external-regulatory audits or certification bodies)
Audits are a key element of a quality management system. The process of establishing an internal audit process can be aided with reference to ISO 19001. This standard provides guidance and lots of examples on implementing and maintaining audit systems.

Benefits of Auditing

Audits provide a means of assessing a company's quality management system, and how well it is in compliance with the processes and procedures within the company. Some key benefits of audits:

- Audits help verify compliance and conformity to requirements laid out in regulations and industry legislation. (e.g. ISO, FDA, Eudralex etc.)
- They measure the effectiveness of the QMS and the engagement of Top Management
- Audit help identify opportunities for improvements

- Audits promotes awareness of the Quality Management System

Quality by Design

The traditional old school of thought focused on 100% inspection where defects are identified by operators. This approach to product quality leads to undetected defects and risk patient safety. This is not to say 100% inspection is not valuable, however, it is more effective is suitable automated systems conduct the inspection.

Online Control or in process control often uses statistical and process controls measures during the manufacturing stage to monitor and respond to drift in process settings and react if defects are produced or detected. The alternative to traditional inspection is to ensure quality is "built in" to the design of a product – Quality by design. Quality by design starts early on within the design and development stage. Potential defects or failure modes are identified and can be designed out of the product or controls put in place to reduce risks or communicate issues.

Quality Management

Quality: Degree to which a set of inherent properties of a product, system or process fulfils requirements.

Risk: defined as the combination of the probability of occurrence of harm and the severity of that harm.

Management: Systematic process for the assessment, control, communication and review of risks to the quality of the drug (medicinal) product across the product lifecycle.

Design Space

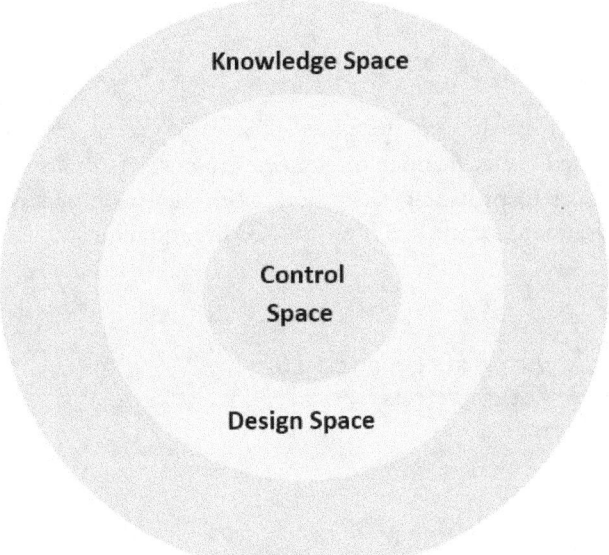

Figure 20 : diagram showing the knowledge space (outer circle) containing both the design space and control space.

Achieving an effective product design, requires in depth knowledge of the customer requirements, clinical or medical need, regulatory requirements, and the manufacturing technology to be used. This collective knowledge or "knowledge space" ensures a robust and quality product is more likely to be designed that will meet the market requirements. Literature, engineering studies and the qualification and experience of employee's all contribute to the knowledge space.

From this knowledge space, the most stable and effective design should be selected with product quality and safety as key factors. Furthermore, the quality of the product is then controlled and maintained within what can be described as the control space.

During the design stage of product development, specifications are created to describe the attributes and features of the product. The output of the design stage is to have the required product specification documents available as inputs to equipment selection and process selection. Examples of some specifications include Raw material specifications, intermediate product specifications and finished product specifications. Specifications contain information on various features and product attributes such as dimensions, formulation, purity, cleanliness, surface finish and so on. The critical requirements stated in specification are often referred to as Critical Quality Attributes (CQAs) and Critical Process Parameters (CPPs)

Critical Quality Attributes (CQA): a particular property of a material, product or output of a process that is key to the product performance and safety.

Critical Process Parameters (CPP): a process parameter such as temperature or time that when varied it impact the quality or CQA of a product.

What Is RFT?

Right First Time strives to create a culture of excellence. People are challenged with performing their tasks always in the correct manner to achieve the correct results always - *right the first time*. RFT is the enabler to providing customers worldwide with accessible, high quality and advanced healthcare solutions which comply with cGMP requirements.

RFT in Practice

Achieve excellence *rather than* "that's good enough"
Prevent defects *rather than* "detect defects"
Right first time *rather than* rework

What is 5S?

5S is a Japanese methodology of organising and storing items in a work or lab environment. It has been adopted by many Western companies as a tool to help maintain standards and reduce errors and mix-ups. The "5s" represents each stage of the method.

Sort
Sorting out any items that are not in use and removing to a more appropriate area or to storage or the bin.

Set-in-Order
The idea of "Set-in-Order" is to be always organised. "A place for everything and everything in its place. "If we "set-in-order" we can help to make live processing and testing more efficient and reduce the risk of errors, omissions and accidents.

Shine
Regular cleaning is an important practice and it is always helpful to "Clean as you go."

Standardise
Implement standard practices through SOPs and training. Standardisation can also be applied to work station layout.

Sustain
Make it a habit! After implementing a 5s methodology, it is only effective if continuous efforts are made to "sustain" the changes.

Sort- Set-in-Order- Shine – Standardise – Sustain

Good Documentation Practices

This section provides an easy-to-understand guide to the subject of Good Documentation Practices. Good Documentation Practices (commonly abbreviated to GDP or GDocP) is a term used to describe standards by which documents are created, modified and maintained. The need for GDP is driven by the general requirement of GMP (Good Manufacturing Practices)

GDP is a practical skill that is required within the life science sector (medical device, pharmaceutical and so on). It can be broadly divided into two streams; GDocP practices and how they apply to electronic document and secondly, GDocP for handwritten entries including initial and dating and recording of data and test results by hand. GDocP is fundamental in achieving compliance to Good Manufacturing Practices (GMP). It is required in the U.S. by the FDA's Code of Federal Regulations and in Europe by the governing body EudraLex. If GDocP is not practiced it jeopardises the integrity of data and written records. This can lead to the falsification of data which is a serious regulatory offence. Admittedly, implementing and maintaining GDP takes time, effort and resources, however, there are some benefits that come with it. Most importantly, Good Documentation Practice is an expected practice in regulated industry as trust and ethics are fundamental to business.

It is important to maintain accurate, complete, up-to-date and consistent information to ensure patient safety and reduce any potential risk to patients. Practicing GDP equally helps to reduce observations raised on inadequate documentation practices at times of audit by regulated bodies such as the FDA. It helps to improve communication and efficiency within companies. If GDP is not followed it can call into question other processes and procedures within a company.

Documentation Creation

The principles of GDP should be applied at the document creation stage. As most people are familiar with softcopy or electronic documents, some of these points are obvious but nonetheless need to be made. All documents should be electronically written and not handwritten except for execution of protocols, test results and adding entries. Documents that are approved controlled should be:

Accurate and free from errors
Have revision or version controlled
Should have an effective date or date of release

Approval of Documents

Document approval must be completed by trained and appropriately experienced personnel. Often companies will use an approval matrix which explains which people are required to approve each document. For example, an EHS (Environment Health and Safety) officer would be required to approve a risk assessment.

Signatures

A signature on any document is legally binding so remember to read and understand what is being signed for. Every signature should also include the date in the correct format. If a signature appears within the same document alongside initials, substituting a full signature with initials and date is generally acceptable. This practice is common when large documents are being completed.

Date and Time Format

A standardised approach to dates and times is important especially within large global organisations. For instance, in the USA, the norm is to place the month before the date, whereas in Ireland and Great Britain it is common to write the day of the month followed by the month. Most companies would define their date and time format in an SOP or procedure.
The date and time format can also be configured in Word documents and Excel worksheets to align with a companies preferred date and time format.

Handwritten Entries

When a handwritten entry is required such as a signature or a test result, indelible ink must be used. Many companies will have an SOP or procedure that states the specific ink colour required. If an entry of a test result or test data isn't completed at the time of execution, this constitutes a late entry. Backdating an entry or signature is forbidden. Always use the correct and current date.

How Are Mistakes Corrected?

This is a critical area of GDP. Failure to follow the requirements of GDP when correcting mistakes is the most common failure when it comes to documentation in industry. The method of correcting mistakes using GDP allows for a person looking at the document for the first time to clearly see the original entry and the corrected entry. This maintains the integrity of the document. In order to identify the changes and corrections, certain rules must be followed. No overwriting is allowed and white-out or Tipp-Ex is not allowed. The correct way to make any changes or corrections by hand is shown in the diagram below.

Accuracy

Accuracy of information provided in documents is critical in the life science industry. As the end user is a patient, inaccurate records or documents could cause serious injury or death. Controlled documents are also legal documents and could be called upon if recalls, litigation or investigations arise.

Many documents used in the manufacture of medical devices are designed to record information or test results. These test results are then used to disposition (pass or fail) batches of product. Inaccurate information could risk the release and distribution of defective product. This has a potential impact on both the business and the patient or user.

Blank Spaces or Blank Fields

On completion of a document such as a logbook or record, no blanks spaces should be left unfilled. This is to avoid late entries and also to prevent confusion. Blank spaces or blank fields should have a diagonal line drawn neatly across the space, the letters "N/A" written and the entry signed and dated. If the reason for "N/A" is not evident then it is wise to include an explanatory note or sentence.

Data Transcription

Transcribing is the process of transferring data from one source to another. This is often required when raw data is involved. When data is in raw format it may need to be entered into a Microsoft Excel sheet. When transcribing data remember that all original raw data must be stored in case it is needed at a future date. After the data is transcribed it must be verified by a second person to check for any errors or omissions.

Revision Control

Controlled documents should always have a version number or revision number electronically on each page of the document. This is similar to books which always list what edition they are e.g. first edition or second edition. Revision control is a key element of the Quality Management Systems in place in regulated industries. As the need for changes in the document arises, the controlled document can be amended/updated. With each update the version number revises also. Some companies will use alphabetic revision control and to a lesser extent numeric revision control (Version A, Version B or Version 01, Version 02).

Management of Attachments

Attachments to controlled documents can include training records, data sheets, lab results and so on. It is important that attachments are identified for traceability purposes. If the attached becomes detached from the main document, the attachment should be identifiable. It is best practice to include a reference number on the attachment if available. If the attachment consists of several pages, each page should be numbered in Page X of Y format if not electronically done so. And remember, hand written entries must be accompanied by a signature and date. Always use staples to attach documents together. Glue or paper clips are not acceptable.

Management of Documents through Their Lifecycle

GDP applies to all the different stages of a document's lifecycle. These stages include creation, review, approval, issuing, completion of records, revision, updating, retirement and storage. Storage a.k.a. retention is an important stage and often a legal requirement for medical devices and pharmaceutical products. For consumer OTC medicines a 5-year retention of quality records often suffices. For implants such as TKRs or Total Knee Replacements, a 90-year retention period is required. This ensures that traceability and a quality record is available if the need arises.

Test Results

This section provides an overview on the correct handling of test results. Test results can be generated from various types of product testing such as visual inspection, dimensional inspection and chemical analysis. The recording of all test results should be completed on an approved form. This is to ensure that the correct information is being recorded and the same approach is taken by different people who might have to complete testing.

Calculations

There are different ways calculations can be completed. Many simple calculations can be done by an individual using a calculator, alternatively, a software package such as Minitab or an Excel sheet can be used to complete complex calculations. It should be clear to the reader what calculation is required, what the formula is and how the calculation is completed.

If the formula used is not included on the sheet, it should be referenced in a controlled document. Care is also required when recording numbers of several decimal places in length, as rounding error can be introduced.

Units of Measurement

The most important thing to remember is consistency in units of measurement when recording data or making calculations. Consult your company procedure if available to determine the correct units of measurement. Many U.S. companies use imperial units e.g. inches, pounds etc. In Europe the International System of Units or SI is used, e.g. millimetres and kilograms.

GLOSSARY

A

Accelerated Aging

When the deterioration of a device or product component from natural aging is accelerated and simulated in the laboratory.

Accuracy

Accuracy or trueness. An expression of the closeness of agreement between the value that is accepted, either as a conventional true value or an accepted reference value and the value obtained. A system with low bias implies good accuracy and vice versa.

Adverse Event

A situation or condition that occurs when a data point, result, or process etc. is outside the expected or predetermined limits or ranges.

Air Exchange Rate Per Hour (ACPH)

The rate of air exchange expressed as number of air changes per hour and calculated by dividing the volume of air delivered in the unit of time by the volume of space.

Active Pharmaceutical Ingredient

Any substance or mixture of substances intended to be used in manufacturing a drug (medicinal) product and that, when used in the production of a drug, becomes an active ingredient of the drug product. Such substances are intended to furnish pharmacological activity or other direct effect in the diagnosis, cure, mitigation, treatment, or prevention of disease, or to affect the structure and function of the body. (ICH Q7A, Annex 18, Part II)

ANSI

American National Standards Institute

Antimicrobial Resistance

Antimicrobial resistance corresponds to the emergence and spread of microbes that are resistant to cheap and effective first-choice, or "first-line" antimicrobial drugs.

Application

A term most often used in relation to Software validation and computerized systems. It is any software installed on a defined platform providing specific functionality.

Approve

"Approve" the device after reviewing a premarket approval (PMA) application that has been submitted to FDA.

AVL (Approved vendor list)
A list of all the vendors or suppliers approved by a company as sources from which to purchase materials.

Artwork

Electronic files or printouts containing the representation of a packaging item, graphical elements, and regulatory text. Approved artworks are used by suppliers for printing.

Aseptic (conditions)

Conditions in the working environment under which the potential for microbial and/or viral contamination is minimized.

ASTM
American society for testing and materials.

ATEX

ATEX, an acronym of the French Atmospheriques Explosives. This European Directive amends and adds safety requirements for hazardous areas in the relevant national legislation in the member states of the European Union, bringing in a common standard. Where equipment is to be used in potentially explosive atmospheres containing gas or combustible dust, it must comply with the ATEX directive.

Audit Trail

The audit trail is a control mechanism of a system that allows all data entered or modified to be traced back to the original data. A reliable and secure audit trail is particularly important in conjunction with the creation, change or deletion of GMP relevant electronic records.

Acceptable Quality Level (AQL)

The AQL of a sampling plan is the Process Performance Level routinely accepted by the sampling plan.

☐

B

Basis of Design

A design document that demonstrates a thorough understanding of the project and its intended output. Typically contains preliminary drawings and system descriptions etc. Together with the URS and the Detailed Design, it provides overall evidence that the design addresses the requirements of the equipment, system or facility.

Biocompatibility

A measure of how a biomaterial interacts in the body with the surrounding cells, tissues and other factors.

Bioburden

The level and type of micro-organisms that can be present in raw materials, API starting materials, intermediates or APIs. Bioburden should not be considered contamination unless the levels have been exceeded or defined objectionable organisms have been detected.

Biological Indicators
Test system containing viable microorganisms providing a defined resistance to a specified sterilization process, e.g. Vaporised hydrogen peroxide.

Biomaterial
Any matter, surface, or construct that interacts with biological systems. Biomaterials can be derived from nature or synthetic (manufactured). The active substance of a biosimilar medicine is comparable to a biological reference medicine. Biosimilar and biological reference medicines are used at the same dose to treat the same disease. The name, appearance and packaging of a biosimilar medicine differs to that of a biological reference medicine.

Bracketing

A Bracketing (aka family or matrix) approach can be used where similar products are produced using the same equipment and processes. A particular product size or product configuration may be selected to represent the worst-case product. Therefore, by qualifying the worst case, all of the other products within the family are considered validated.

Body orifice

Any natural opening in the body, as well as the external surface of the eyeball, or any permanent artificial opening, such as a stoma or permanent tracheotomy.

Borderline Classifications

In certain circumstances, it may not be clear if a product falls under the medical device legislation or whether to classify a device as a medicine, cosmetic, biocide and so on. The decision will largely depend on the particular intended use of the product, as assigned by the manufacturer, and on the demonstrated mode of action. The manufacturer's claims must be substantiated by relevant data.

Bulk Product

Any pharmaceutical form (liquid, powder, suspension) that is to be filled into either another container or its final container at the next process step; or is already filled into its final container to be labelled and packaged at the next process step.

BOM

Bill of Materials.

BSI

British Standards Institute.

C

CAD, Computer Aided Drawing

A system used to create physical designs, usually three-dimensional. Some examples of CAD software are SolidWorks, Pro/ENGINEER and AutoCAD.

Calibration

The a requirement that demonstrates a particular instrument or device produces results within specified limits by comparison with those produced by a reference or traceable standard over an appropriate range of measurements.

Campaign (Process)

A production strategy where consecutive batches of an API, a finished product, or intermediates are processed before the production line/system is cleaned.

Capability (Process Capability)

Process Capability is a measure of how capable the process is of producing product meeting specified requirements. It is a measure of the actual variation in that product characteristic compared to the product specifications. Indices are used to represent the Process Capability such as Pp, Cp and Ppk, Cpk, depending on how the data is collected e.g. multiple batches over time.

CAPA

A Corrective and preventive action. A systematic approach that includes actions needed to correct, prevent recurrence and eliminate the cause of potential nonconforming product and other quality problems (preventive action) (21CFR 820.100)

Change Control

A formal system by which qualified representatives of appropriate disciplines review proposed or actual changes that may impact the validated status.

Change Notification (Agreement)

A signed declaration that states that the Supplier agrees to notify the customer of changes in its product or process in order to allow the customer determine whether the changes can affect the quality of finished goods or Quality System.

Change Management

An overarching approach to change control that is used during the preliminary planning and design stage of a project.

Cleaning

The process of removing potential contaminants from process equipment and maintaining the condition of equipment such that the equipment can be safely used for subsequent product manufacture.

Cleaning Validation

Documented evidence that provides a high degree of assurance that a specific cleaning process will consistently produce a result meeting predetermined requirements for cleanliness.

Cleaning Verification

Confirmation by examination and provision of objective evidence that specific requirements have been fulfilled.

Cocurrent (flow)

This is when the fluids are applied in the same direction. Cocurrent flow is less effective as less heat can be transferred, therefore it is less commonly used.

Code of Federal Regulations (CFR)

Regulations issued by U.S. government agencies. The individual titles making up the regulations are numbered the same way as the federal laws on the same topic.

Competent Authority

A competent authority is the legally designated authority mandated to monitor compliance with directives and legal requirements within the industry. The competent authority has the power to grant and revoke licenses.

Compendial Organisations

Organizations certifying material standards that meet compendial requirements and acceptance criteria. (e.g. USP).

Commissioning

An engineering activity that includes all aspects of bringing a system, piece of equipment or process is installed and ready for use. Commissioning involves both requirements of Installation Qualification (IQ) and Operational Qualification (OQ).

Computer System

A group of hardware components and associated software, designed and assembled to perform a specific function or group of functions.
[EU GMP Guide, Part II, ICH Q7]

Computerised System

A system including the input of data, electronic processing and the output of information to be used either for reporting or automatic control. [EU GMP Guide, Glossary]

Computer System Validation

A process that confirms by examination and provision of objective evidence that the computer system conforms to user needs and intended uses. System validation is a process for achieving and maintaining compliance with GxP regulations and fitness for intended use by adoption of life cycle activities, deliverables, and controls.

Concurrent Validation

Concurrent Validation occurs when activities are executed at the same time as one another or concurrent to a product launch.

Confidence Level

Confidence Level is expressed as a percentage and represents the probability that the conclusion of the test is correct. A 95% confidence level means you can be 95% certain that the conclusion is correct.

Conflict Of Interest

A conflict of interest is a situation in which a public official's decisions are influenced by the official's personal interests.

Continual Improvement, CI

Ongoing activities to evaluate and positively change products, processes, and the quality system to increase effectiveness

Consent Decree

A consent decree is a binding order issued by a judge that stipulates the voluntary agreement by the participants in a case of litigation. Decrees are sometimes issued after one party voluntarily agrees to cease a particular action without admitting to any illegality of the action to date.

Colony Forming Unit

One or more microorganisms that produce a visible, discrete growth on an agar-based microbiological medium.

Controlled Substances

Products that are categorized due to their potential for abuse, medical use and requirement for medical supervision.

Controlled classified areas

An environment supplied with HEPA filtered air where materials, equipment, and personnel are regulated to control viable and non-viable particulates to an acceptably low level. Such areas are classified according to the maximum level of airborne particulate allowed.

CNC (Controlled Not Classified)

While these are not ISO recognized room classes, they are generally used to describe non-GMP areas with a level of control in effect.

Clear (FDA)

"clear" the device after reviewing a premarket notification, otherwise known as a 510(k) (named for a section in the Food, Drug, and Cosmetic Act), that has been filed with FDA, or

Cleanroom

An area (or room or zone) with defined environmental control of particulate and microbial contamination, constructed and used in such a way as to reduce the introduction, generation and retention of contaminants within the area.

Containment

A process or device to contain product, dust or contaminants in one zone, preventing it from escaping to another zone.

Contamination

The undesired introduction of impurities of a chemical or microbial nature, or of foreign matter, into or onto a starting material or intermediate, during production, sampling, packaging or repackaging, storage or transport.

Continued Process Verification

Once the initial validation is completed it is important that the system or process remains within the validated state. This is done by monitoring the performance and output of the system or equipment. Furthermore, any changes to this system or equipment must be assessed and documented in order to assure the product is safe and meets acceptance criteria.

Critical Aspects

Critical aspects of manufacturing systems include the functions, features, abilities, and performance or characteristics required for the manufacturing process and systems to ensure consistent product quality and patient safety. They should be identified and documented based on scientific product and process understanding.

Critical Quality Attribute, CQA (Critical-to-Quality)

A property or characteristic with specific nominal value and appropriate limit and range providing a particular quality attribute. A CQA typically is classed as a high risk requirement, where the safety or efficacy of the product depends on the CQA been within the specified limits.

CCC (Mark)

The "China Compulsory Certificate"mark, commonly known as CCC Mark, is a safety mark for many products sold on the Chinese market. As of 2013, medical devices do not require this certification.

CDC

Center for Disease Control & Prevention (USA)

CDRH

Center for Devices and Radiological Health (USA)

CE Marking

The CE Marking is a mandatory conformance mark on many products (including medical devices) placed on the single market in the European Economic Area. The CE marking certifies that a product has met EU consumer safety, health or environmental equirements. By affixing the CE marking to a product, the manufacturer declares that it meets EU safety, health and environmental requirements.

CEN
Communité Européenne des Normes (European Committee for Standardization).

Clinical Trial

Clinical Trials are conducted to allow safety and efficacy data to be collected for health interventions (e.g., drugs, diagnostics, devices, therapy protocols). These trials can take place only after satisfactory information has been gathered on the quality of the non-clinical safety, and Health Authority/Ethics Committee approval is granted in the country where the trial is taking place.

Clinical Trial Sponsor

The Clinical Trial Sponsor is responsible for the safety of subjects in a clinical trial and informs local site investigators of the true historical safety record of the drug, device or other medical treatment to be tested, and of any potential interactions of the study treatment(s) with already approved medical treatments.

Cleaning

Removal of contamination or soils from an item or surface to the extent necessary for its further processing and its intended subsequent use.

CMDCAS

Canadian Medical Devices Conformity Assessment System.

CMDR

Canadian Medical Device Regulation.

Conformity

Fulfilment of a requirement or meeting a requirement.

Conformity Assessment Body (CAB)

A body, other than a Regulatory (competent) Authority, engaged in determining whether the relevant requirements in technical regulations or standards are fulfilled.

CRO

A "Contract Research Organization", also commonly known as a "Clinical Research Organization", is a service organization that provides support to the pharmaceutical and biotechnology industries. CROs offer clients a wide range of "outsourced" pharmaceutical research services to aid in the drug and medical device research and development process.

D

Data Integrity

Is the degree to which data is reliable and without error. Data must be accurate, attributable, contemporaneous, original, legible and available. A breach of data integrity occurs when any person manipulates or distorts data and submits the results of that data as valid.

Dead Leg

A dead leg in the world of piping terminology refers to an area of piping where there is insufficient flow or a tendency for water build-up or stagnation.
The formal definition of a dead-leg states that
Pipelines for the transmission of purified water for manufacturing or final rinse should not have an unused portion greater in length than 6 diameters (6D rule) of the unused portion of pipe measured from the axis of the pipe in use.

Debugging

The process of locating, analysing, and correcting suspected faults or machine issues. 333

Design controls

Design controls are a collection of practices and procedures that are incorporated into the design and development process for a product such as a medical device. It provides a structure and clear path from user needs assessment to product delivery through a step-by-step process. Design controls ensure proper assessment of the design is completed during the design and development phase. Design controls are a requirement of quality systems such as 21 CFR Part 820 (medical devices), and for certain classes of devices and per ISO 13485 - Quality Management Systems.

Decommissioning

When a system is taken out of production service and stored in an adequate environment for potential future use.

Depyrogenation

A thermal process used to destroy or remove pyrogens (endotoxins). Typically primary packaging components such as glass vials are subject to Depyrogenation.

Detection Limit

The lowest amount of analyte in a sample that can be detected but not necessarily quantitated as an exact value for an individual analytical procedure. (Ref: ICH Q2)

Design History File

The DHF is a repository for all of the documentation generated as a result of the design control process. The DHF serves as a complete record of the design.

Design Validation

Establishing by objective evidence that device or product specifications conform to user needs and intended use(s) defined in design documentation.

Debarment

The FDA has the authority to "disqualify," or remove, researchers from conducting clinical testing of new drugs and devices when the agency determines that the researcher has repeatedly or deliberately not followed the rules intended to protect study subjects and ensure data integrity. Further, the FDA can disqualify a clinical investigator who has repeatedly or deliberately submitted false information to the agency or study sponsor in a required report.
Under its statutory debarment authority, the agency may also ban, or "debar" from the drug industry individuals and companies convicted of certain felonies or misdemeanours related to drug products. Once individuals have been subjected to "debarment," they may no longer work for anyone with an approved or pending drug product application at FDA. Debarred companies may no longer submit abbreviated drug applications.

Design qualification (DQ)

The documented verification that the proposed design of the is suitable for the intended purpose. DQs are typical deliverables for facilities, systems and equipment and or processes.

Design Space

The multidimensional combination and interaction of input variables, e.g. material attributes, and process parameters that have been demonstrated to provide assurance of quality. Working within the design space is not considered as a change.

Directives

Directives are legal requirements. These must be met by manufacturers. Standard such as ISO 13485 help companies meet the requirements of directives, such as "Guidelines Relating to the Application of the Council Directive 93/42/EEC on Medical Devices."

Direct impact (system)

A system that is expected to have a direct impact on product quality. These systems are designed and commissioned in line with Good Engineering Practice (GEP) and, in addition, are subject to Qualification and Validation. Such systems include HVACs and Clean utilities such as WFI (Water-for-Injection)

Diffusion blending

A process in which particles are reoriented in relation to one another when they are placed in random motion and interparticular friction is reduced as the result of bed expansion (usually within a rotating container). Also referred to as tumble blending.

Deviations

A deviation can be simply described as an unintended event which causes a test or verification to fail to meet expected acceptance criteria.

Degree of invasiveness

A device, which in whole or in part, penetrates inside the body either through a body orifice or through the skin surface, is invasive. Invasiveness is generally categorised as invasive of a body orifice (including the surface of the eye), surgically invasive devices and implantable devices.

Device Master Record (DMR)

a compilation of records containing the procedures and specification for a device. The contents of a DMR can contain local procedures such as SOPs and work instructions along with global or divisional specifications used to detail manufacturing processes, intermediate product or final product.

Drug Product

The dosage form in the final immediate packaging intended for marketing. The finished dosage form that contains a Drug Substance, generally, but not necessarily in association with other active or inactive ingredients. (FDA)

Duration of Contact

In determining the classification of a device the duration that the device is in continuous contact with the patient is defined as transient, short term or long term. The longer the device is in contact with the patient or user, the greater the risk and therefore this has to be taken into account when determining classification. Continuous use is defined in MEDDEV 2.4/1 as the uninterrupted actual use for the intended purpose. Where use of a device is discontinued in order that the device is immediately replaced with an identical device (e.g. replacement of a urethral catheter) this shall be considered as continuous use of the device.

E

Electronic Signatures

Electronic signatures are computer-generated character strings that count as the legal equivalent of a handwritten signature. The regulations for the use of electronic signatures are set out in 21 CFR Part 11 of the FDA. Each electronic signature must be assigned uniquely to one person and must not be used by any other person. It must be possible to confirm to the authorities that an electronic signature represents the legal equivalent of a handwritten signature. Electronic signatures can be biometrically based or the system can be set up without biometric features.

Encapsulation

The division of material into a hard gelatin capsule. Encapsulators should all have the following operating principles in common: rectification (orientation of the hard gelatin capsules), separation of capsule caps from bodies, dosing of fill material/formulation, rejoining of caps and bodies, and ejection of filled capsules.

Endotoxin

A pyrogenic product (e.g., lipopolysaccharide) present in the bacterial cell wall. Endotoxin can lead to reactions in patients receiving injections ranging from severe fever to death.

Equipment Qualification

Qualification means the process to demonstrate the ability to fulfil specified requirements. EQ consists of proving and documenting that equipment or ancillary systems are properly installed (Installation Qualification, IQ), work correctly (Operations Qualification OQ), and the different sub-systems work together as a system (Performance Qualification PQ) and actually lead to the expected results.
Qualification is part of validation, but the individual qualification steps alone do not constitute a validated process.

Excipient

Substances other than the API which have been appropriately evaluated for safety and are intentionally included in a drug delivery system to provide a specific role in manufacturing, shelf-life or physical property.

Equipment Range

The full range that equipment is capable of performing, as per the manufacturer specification and tolerances. (a process may not utilize the full equipment range, operating over a narrower range).

F

Factory Acceptance Testing (FAT)

An FAT or Factory Acceptance Test is an engineering activity that inspects and verifies that the equipment or system meets the requirements of the URS.

Failure Mode And Effects Analysis (FMEA):

A risk assessment tool that provides for an evaluation of potential failure modes and their likely effect on outcomes and/or product or process performance in order to prioritize risks and monitor the effectiveness of risk control activities. It is often used to identify areas within a given process, product, or system that render it vulnerable.

FDA 483s

An FDA 483 letter typically includes a summary of findings and observations in relation to an audit or inspection where the FDA representatives have reason to believe GMP or other regulations have been violated or are not being met. In response to an FDA 483 letter, the company should address each item and provide a timeline for correction or request clarification of what changes are required.

Functional Design Specification (FDS)

A functional design specification is a document that specifies how particular requirements are met – this can be a combination of how the equipment/process operates mechanically/automatically etc. An FDS is typically written to response to a URS

Fluid

A fluid is a substance that undergoes continuous deformation when subjected to a shearing force.

☐

☐

G

GAMP

Good Automated Manufacturing Practice (GAMP) is a set of guidelines for manufacturers and users of automated systems in regulated industries. Specifically, the Medical device, pharmaceutical and biopharmaceutical industries.
The application of GAMP and Validation of Automated Systems in manufacturing helps ensure that regulated medical devices and medicinal products have the required quality and are manufactured according to Good practices, meet regulatory and legal requirements and ensure patient safety.

Good Documentation Practices, GDP

The handling of written or pictorial information describing, defining, specifying and/or reporting of certifying activities, requirements, procedures or results in such a way as to ensure data integrity

Granulation

A process of creating granules. The powder morphology is modified through the use of either a liquid that causes particles to bind through capillary forces or dry compaction forces.

Grade A Areas

Aseptic processing areas, critical in nature where sterile products are exposed to the environment receiving no further sterilization. High-risk operations (for example aseptic stopperage, filling, loading of the lyophilizer) occur in Grade A areas. They are considered ISO 5 under both dynamic and static conditions.)

Grade B Areas

Aseptic processing areas where the sterile product is protected from the environment. Grade B processing areas are the background environments for Grade A areas and are considered ISO 7 environments in the dynamic state and ISO 5 environments under static conditions.

Grade C Areas

Non-critical areas where bulk product or materials are exposed to the environment, yet final sterilization has not yet been performed. Grade C areas are support areas for non-sterile production activities; purification, formulation, and preparation of components, equipment, etc. for sterilization. They are considered ISO 8 (Class 100,000) environments in the dynamic state and ISO 7 (Class 10,000) environments under static conditions.

Grade D Areas

Non-critical production areas, support areas, airlocks, or corridors. They are support areas for non-sterile production activities in closed systems; cell culture, or buffer and media preparation areas. Grade D Airlocks are used for the movement of product, materials, and personnel into classified areas.

GHTF

Global Harmonization Task Force

GxP

GxP is a general term for good practice with regard to quality guidelines and regulations. These guidelines are used in many fields, including the pharmaceutical, medical device and food industries. "x" is used as an umbrella letter representing different subjects or disciplines in industry. Some prime examples include GLP (Good Laboratory Practice), GDP (Good Documentation Practice), GEP (Good Engineering Practice) and GMP (Good Manufacturing Practices). Furthermore, the use of a lower case "c" as a prefix indicates "current" or "up-to-date"

☐

H

Harm

Damage to health, including the damage that can occur from loss of product quality or availability.

High level risk assessment (HLRA)

A High level risk assessment that can be used at the beginning of a project to estimate the risk. Such as the risks involved with bringing in new computerised/automated equipment.

HVAC

Heating, ventilation and air-conditioning (HVAC) systems are used to control the environmental conditions within an area or manufacturing facility. HVAC systems also provide comfortable conditions for operators based in the manufacturing environment. Temperature, relative humidity (RH) and ventilation should not adversely affect the quality of products during their manufacture and storage, or the proper functioning of equipment

Hydrogel

A biomaterial made up of a network of polymer chains that are highly absorbent and as flexible as natural tissue.

☐

I

ICH

International Conference on Harmonization of Technical Requirements for Registration of Pharmaceuticals for Human Use.

Intended Purpose

Intended purpose means the use for which the device is intended according to the data supplied by the manufacturer on the labelling, in the instructions and/or in promotional materials. (Chapter I section 1 of Annex IX of Directive 93/42/EEC)

Impurity

Any component of the new active pharmaceutical ingredient which is not the chemical entity defined as the new active pharmaceutical ingredient OR any component present in the active pharmaceutical ingredient or final product which is not the desired product, a product-related substance, or excipient including buffer components.

Invasive device

A device, which, in whole or in part, penetrates inside the body, either through a body orifice or through the surface of the body.

IQ/OQ

Equipment IQ/OQ is defined as establishing documented evidence that all key aspects of the process equipment installation adhere to the manufacturer's approved specifications and any recommendations of the supplier of the equipment are suitably considered.
The process/equipment must also operate as intended and all user requirements are adequately fulfilled.

IFU

Instructions for Use.

(Plant) Injunction

An injunction is a judicial process initiated to stop or prevent violation of the law, such as to halt the flow of violative products in interstate commerce and to correct the conditions that caused the violation to occur. (FDA 21 U.S.C. 332; Rule 65, Rules of Civil Procedure).
If a firm has a history of violations and has promised correction in the past but has not made the corrections, the injunction is more likely to succeed. However, the freshness of the evidence is critical.

For an injunction action to be credible in the eyes of the Department of Justice (DOJ), the U.S. Attorney and the court, the evidence must be current. Timeliness is an important factor when considering an injunction action, with or without a Motion for Preliminary Injunction or a temporary restraining order (TRO). However, case quality and credibility must not be sacrificed to meet guideline time frames. The purpose of the guideline time frames is to limit, as much as can reasonably be expected, the need to update evidence. Updating entails extra work at all levels of the case development and review process and more importantly, delays obtaining an injunction which is intended to stop violations that adversely affect the safety or quality of products in commerce.

ISO

International Organization for Standardization. Agency responsible for developing international standards. E.g. ISO 13485 Medical Devices.

Isolator
A sealed enclosure, which provides full physical separation between the critical processing zone and the surrounding other processing zones. The internal surfaces of the isolator and of its contents are decontaminated, in accordance with defined objectives, by highly effective cycles. (e.g. Vaporised Hydrogen peroxide) Enclosure capable of preventing ingress of contaminants by means of physical interior/exterior separation, and capable of being subject to reproducible interior bio-decontamination.

Isoelectric Precipitation

Isoelectric Precipitation works by reducing the electrostatic forces to near zero, allowing the proteins to precipitate out.

ISO 13485

ISO 13485, ISO standard, published in 2003, that represents the requirements for a comprehensive management system for the design and manufacture of medical devices.

ISO 14971

An ISO standard, published in 2007, that provides a framework and requirements for a risk management system for medical devices. This standard establishes the requirements for risk management to determine the safety of a medical device by the manufacturer during the product life cycle.

ISO 9001

ISO 9001 is an ISO standard that represents the requirements for quality management systems. It is used across industries and is not specific to medical devices like ISO 13485.

Item Master

A of all components that a manufacturer buys, builds or assembles into its products. The item master includes information like the size, shape, material, manufacturer, manufacturer part number and vendor for each component.

IVD

In Vitro Diagnostic tests are medical devices intended to perform diagnoses from assays in a test tube, or more generally in a controlled environment outside a living organism.

IVDD

The In Vitro Diagnostic Device Directive delineates requirements that in vitro diagnostic devices must meet before they can be sold in the EU market.

Intermediate

A material produced during steps of the processing of an API that undergoes further molecular change(s) or purification before it becomes an API.

J

JIT (Just in time)

A strategy used to monitor inventory levels with the goal of reducing inventory and associated carrying costs.

K

Kanban

A scheduling system that advises manufacturers what to produce, when to produce and how much to produce. Pioneered by Toyota, the approach is based on demand. Inventory is replenished only when visual cues like an empty bin, trolley or cart show that it's needed.
☐

L

Laminar flow

Laminar flow is when fluid particles move in parallel layers, at a constant velocity.

Lifecycle (Validation)

The Validation lifecycle refers to the requirement to control and document all validation activities from conception and URS stage to the retirement of equipment or a process. The lifecycle approach ensures compliance throughout the life of the process/equipment while maintaining a validated state throughout the application of change control.

Linearity

The ability of an analytical procedure (within a given range) to obtain test results that are directly proportional to the concentration (amount) of analyte in the sample.

Line Clearance

The act of performing and documenting the removal of materials from a production or packaging line and cleaning prior to the introduction of a new batch or lot.

Lyophilization (or Freeze Drying)

Lyophilization is the removal of ice or other frozen solvents from a material through the process of sublimation and the removal of bound water molecules through the process of desorption.

M

Maximum Allowable Carry Over (MACO)

The amount of allowed product residue (carry-over) from lot-to-lot, batch-to-batch. This limit is based on the most conservative or lowest level of three MACO calculation methods (1) Limited based on Toxicity, (2) Limit based on Smallest Therapeutic Dose, and (3) Worst Case Dose.

Measurement Capability Index (MCI)

The Measurement Capability Index (MCI) represents the capability of the measurement system. It is used to evaluate the capability of the gauge to classify product against predetermined specifications.

Measurement System Analysis (MSA)

A study to determine the degree of error involved in measuring the given parameter. The measurement system involves the combination of operations, procedures, gauges, instruments, environmental conditions, people and software.

Medical Device

A medical device is "an instrument, apparatus, implement, machine, contrivance, implant, in vitro reagent, or other similar or related article, including a component part, or accessory which is:

- recognized in the official National Formulary, or the United States Pharmacopoeia, or any supplement to them,
- intended for use in the diagnosis of disease or other conditions, or in the cure, mitigation, treatment, or prevention of disease, in man or other animals, or
- intended to affect the structure or any function of the body of man or other animals, and which does not achieve any of its primary intended purposes through chemical action within or on the body of man or other animals and which is not dependent upon being metabolized for the achievement of any of its primary intended purposes."

Medicinal Drug Products (Finished Products)

Finished dosage forms (e.g. tablet, capsule, or solution) that contain the active pharmaceutical ingredient usually combined with inactive ingredients. Medicinal products are intended to furnish pharmacological activity or other direct effect in the diagnosis, cure, mitigation, treatment, or prevention of disease or to affect the structure and function of the body.

MDD

The Medical Device Directive is intended to harmonize the laws relating to medical devices within the European Union. Medical Device Directive 93/42/EEC was most recently reviewed and amended by 2007/47/EC.

MHRA

The Medicines and Healthcare products regulatory Agency (MHRA) is the UK government agency which is responsible for ensuring that medicines and medical devices work and are acceptably safe.

MSDS

Material Safety Data Sheet.

N

NCR

Non-Conformance Report.

NIH

National Institutes of Health (U.S.)

Noel

No Observed Effect Level. In relation to Cleaning Validation.

Non-conformity

A deficiency in a characteristic, product specification, CQA, process parameter, record, or procedure that renders the quality of a product unacceptable, indeterminate, or not according to specified requirements.

Non Parametric Data

Where the type of data is non variable Also referred to as attribute data eg (Visual inspection resulting in a PASS/FAIL result.

Notified Bodies

A notified body is a certification organisation which the national authority (the competent authority) of a member state designates to carry out one or more of the conformity assessment procedures or audits described in the annexes of the medical devices directives or GMP legislation.

NPI (New product introduction)

The market launch or commercialization of a new product. NPI takes place at the end of a successful product development project.

☐

O

Open System

An environment in which system access is not controlled by persons who are responsible for the content of electronic records on the system (21 CFR, Part 11)

Outlier

A test result that is statistically different compared to a set of other test results obtained from the same sample or samples from the same lot of material.

Out-Of-Specification

A recorded result that falls outside the established specification(s) or acceptance criteria.

Out-Of-Trend

Analytical result, which is within specification or acceptance criteria, but different from those usually obtained or expected. Out-of-trend results should be investigated by the same general principles as out-of-specification results.

Quantitation limit

The lowest amount of analyte in a sample which can be quantitatively determined with suitable precision and accuracy for an analytical procedure. The quantitation limit is a parameter of quantitative assays for low levels of compounds in sample matrices and is used particularly for the determination of impurities and degradation products.

Overall Equipment Effectiveness(OEE)

A calculation for measuring the efficiency and effectiveness of a process, by Equipment breaking it down into three constituent components (the OEE Factors) Availability x Performance x Quality.

Overkill

Sterilization process that is demonstrated as delivering at least a 12 Spore Log Reduction (SLR) to a biological indicator having a resistance equal to or greater than the bioburden level.

P

Pan Coating

The uniform deposition of coating material onto the surface of a solid dosage form while being translated via a rotating vessel.

Particle count test

Test covers verification of cleanliness. Dust particle counts measured. The number of readings and positions of tests should be defined in accordance with ISO 14644-1 Annex B5.

Performance indicators

Measurable values used to quantify quality objectives to reflect the performance of an organization, process or system, also known as performance metrics in some regions. (ICH Q10)

Performance Qualification (PQ)

Establishing by documented evidence that the process, under anticipated (controlled) conditions, consistently produces a product which meets predetermined requirements.

Precision

The degree of agreement (scatter) between a series of measurements when a method is applied repeatedly to multiple samplings of a homogeneous sample or artificially prepared sample under the prescribed conditions. There are three types of precision; repeatability, intermediate precision and reproducibility.

Pressure cascade

A process whereby air flows from one area, which is maintained at a higher pressure, to another area at a lower pressure.

Piping & Instrument Diagrams (P&IDs)

Engineering technical drawings that provide details of the connections and integration of equipment, services, material flows, plant controls and alarms. The P&ID also provide the reference for each tag or label used for identification.

PMA

Premarket approval by FDA is the required process of scientific review to guarantee safety and effectiveness for Class III devices.

PMDA

The Pharmaceutical and Medical Devices Agency in Japan reviews applications for marketing approval of pharmaceuticals and medical devices. It also monitors their post-marketing safety and provides relief compensation for people who have suffered from adverse drug reactions from pharmaceuticals or infections from biological products.

PMS

Post Marketing Surveillance is the practice of monitoring a pharmaceutical drug or device after it has been released on the market.

Process design

Defining the commercial manufacturing process based on knowledge gained through development and scale-up activities.

Process qualification

Confirming that the manufacturing process as designed is capable of reproducible commercial manufacturing.

Process window

The selected operating range of machine setting/parameter that will produce product to meet all quality and product specifications.

Product Recovery

Product recovery is a critical and important step in the process. It is also referred to as "Downstream processing". It is often the most expensive step in the process. For recombinant-DNA derived products, purification can often account for 90% of the total production costs.

Prospective Validation

Prospective Validation is when validation is done in advance of commercial manufacturing.

Procedures

Also known as Standard Operating Procedures, or SOPs, give directions for performing certain operations.

Protocols

Give instructions for performing and recording certain discreet operations. (Examples include engineering protocols, validation protocols etc.)

Pure

A term typically used within pharmaceutical manufacturing, a product or substance is pure if it is free of contaminants, foreign matter, chemicals and harmful microbes.

Q

QMS

Quality Management System can be expressed as the organizational structure, procedures, processes and resources needed to implement quality management.

Quality

The degree to which a set of inherent properties of a product, system, or process fulfils requirements. (ICH Q9)

Quality by design

This is a systematic approach that begins with predefined objectives and emphasizes product and process understanding and process control, based on sound Science and engineering principles.

Quality Management System

A Quality Management System, often abbreviated to (QMS) is any system based on a collection of business processes that are primarily focused on providing safe and quality products that consistently meet customer requirements.

Quarantine

The status of materials isolated physically or by other effective means pending a decision on their subsequent approval or rejection.

(Quality) Policy

A document in which a company or organization outlines their commitment and approach to quality. It usually sets out how they plan to achieve a high and consistent standard of quality. It should in some way speak to the customer or end user.

Qualification Plan

A Qualification Plan (QP) describes all the qualification measures and at which stage of the qualification the verification will be completed. It typically contains detailed descriptions of the necessary test measures and a description of the interdependencies of the individual tests. In some instances, there may not be a need or a requirement for a qualification plan. A validation plan can also serve to detail the qualification strategy.

QP

Companies that intend to manufacture or import medicinal products or intermediate products, for use in clinical trials or for market within the EU, must appoint the service of a Qualified Person, in order to comply with EU Good Manufacturing Practice Standards.

QPM

Quality Policy Manual.

QSP

Quality System Procedure.

QSR

Quality System Regulations.

R

Range

Range is defined as the interval between the upper and lower measurements required. The minimum specified range should be within the equipment range and validated to operate at all points within the range.

Recall

As defined at 21 CFR 7.3(g), "recall means a firm's removal or correction of a marketed product that the Food and Drug Administration considers to be in violation of the laws it administers and against which the agency would initiate legal action, 2 21 CFR 806.2(h). e.g., seizure. Recall does not include a market withdrawal or a stock recovery." Recall does not include routine servicing. Recall also does not include an enhancement, as defined by this guidance.

Relative humidity

The ratio of the actual water vapour pressure of the air to the saturated water vapour pressure of the air at the same temperature expressed as a percentage. More simply put, it is the ratio of the mass of moisture in the air, relative to the mass at 100% moisture saturation, at a given temperature.

Reusable medical device

A device intended for repeated use either on the same or different patients, with appropriate decontamination and other reprocessing prior to re-use.

Reusable Surgical Instrument

Instrument intended for surgical use by cutting, drilling, sawing, scratching, scraping, clamping, retracting, clipping or similar surgical procedures, without connection to any active medical device and which are intended by the manufacturer to be reused after appropriate procedures for cleaning and/or sterilisation have been carried out.
☐
Re-Qualification

Requalification is designed to verify and ensure that the equipment/instrument/system is maintained in a qualified state after modification or after a stipulated time period (downtime).

Residual Risk

The risk level remaining after applying the identified controls on a high risk of harms and hazards manifestation.

Resolution

The smallest change in quantity that can be detected or provided by an instrument.

Residual Solvent
Organic volatile chemicals used or produced during the manufacture of APIs or excipients, or in the preparation of medicinal products.

Retain Samples
Samples that are kept for potential investigations and retests. It should be noted that retain samples are not a regulatory requirement, per Annex 10 or 21 CFR part 11.

Retrospective Validation

Retrospective validation is used for facilities or processes that have not completed formal validation. Historical data or a retrospective review can provide the evidence that the process or facility is operated as intended.

Rinse Sampling

Using a solvent to contact all surfaces of the sampled item to quantitatively remove target residue. The solvent can be water, water with pH adjusted, or organic solvent.

Right First Time

Right First Time strives to create a culture of excellence. People are challenged with performing their tasks always in the correct manner to achieve the correct results always - right the first time.

Risk

The combination of the probability of occurrence of harm and the severity of that harm.

Risk Management

Risk management involves the systematic application of management policies, practices and procedures that identify, analyse, control and monitor risk.
It is important to recognise that risk management should begin at the outset of the design and development phase of a project. The first step is to identify the user needs and intended use and application of the device.

RoHS

"Restriction of Hazardous Substances in electrical and electronic equipment 2002/95/EC". An initiative that was adopted by the European Union (EU) in February 2003 and put into effect July 1, 2006
Ruggedness

An indication of how resistant a test method or process is to typical variations in operation, such as those to be expected when using different analysts, different instruments and different reagent batches.
☐
☐

S

Scaffold

A structure of artificial or natural materials on which tissue is grown to mimic a biological process outside the body.

SKU

(Stock keeping unit) A unique sales stock identifier.

Specifications

A approved document detailing the requirements with which the products or materials used or obtained during manufacture have to conform. They serve as a basis for quality evaluation.

Specificity

The ability to assess unequivocally the analyte in the presence of components, which may be expected to be present.

Stability

Stability studies are used to demonstrate and justify assigned expiration or retest dates.

5S

5S is a Japanese methodology of organising and storing items in a work or lab environment. It has been adopted by many Western companies as a tool to help maintain standards and reduce errors and mix-ups. The "5s" represents each stage of the method.
Sort
Sorting out any items that are not in use and removing to a more appropriate area or to storage or the bin.
Set-in-Order
The idea of "Set-in-Order" is to be always organised. "A place for everything and everything in its place. "If we "set-in-order" we can help to make live processing and testing more efficient and reduce the risk of errors, omissions and accidents.
Shine
Regular cleaning is an important practice and it is always helpful to "Clean as you go."
Standardise
Implement standard practices through SOPs and training. Standardisation can also be applied to work station layout.
Sustain
Make it a habit! After implementing a 5s methodology, it is only effective if continuous efforts are made to "sustain" the changes.

Sterility Assurance (SAL)

Probability of a single viable microorganism occurring on an item after sterilization. For a terminally sterilized medical device to be designated as "sterile", the minimum sterility assurance level shall be SAL = 10-6 or better. When applying this quantitative value to assurance of sterility, an SAL of 10-7 has a lower value but provides a greater assurance of sterility than an SAL of 10-6 .

☐

T

Tableting

The reconstitution of a powder blend in which compression force is applied to form a single unit dose. (tablet)

Tableting press

Tablet press subclasses primarily are distinguished from one another by the method that the powder blend is delivered to the die cavity. Tablet presses can deliver powders without mechanical assistance (gravity), with mechanical assistance (automation), by rotational forces (centrifugal), and in two different locations where a tablet core is formed and subsequently an outer layer of coating material is applied (compression coating).

Traceability Matrix

A Traceability Matrix is a document that links the user requirements and specifications to where the verification and testing has been documented within the validation activities.
A traceability matrix illustrates that all user requirements are traceable to the evidence based test.

Turbulent flow

Turbulent flow is when the movement of fluid particles are varying in velocity and direction.

U

Uniform

The product is manufactured consistently and will have the same quality between batches manufactured on different days.

UDI, Unique Device Identification

The UDI is a series of numeric or alphanumeric characters that is created through a globally accepted device identification and coding standard. It allows the unambiguous identification of a specific medical device on the market.

Uninterrupted Power Supply

An uninterruptible power supply (UPS) is a system for buffering the main power supply. If the power supply fails, the battery of the UPS supplies the required power. When the power supply returns, the UPS battery stops supplying power and is recharged.

Unit Operation

Unit operations are the individual steps in the process that modify materials and their properties at each step of the process. Each unit operation comes together to create a complete process.

User Requirement Specification, URS

The URS is a critical document that defines the requirements of a particular system, equipment or process. Requirements such as the functional and operational aspects of the system are typically documented here.

USP

United States Pharmacopoeia.

V

Validation

Validation is confirmation via documented evidence that the particular requirements for a specific intended use can be consistently fulfilled under anticipated conditions.

Validation Master Plan

A document providing information on a company's validation work programme. It typically details timescales for the validation work to be performed along with the key deliverables.

Verification

Verification confirmation by examination and provision of objective evidence (i.e. documentation) that the specified requirements have been fulfilled.

Vaporized Hydrogen Peroxide (VHP)

 Vaporization of liquid hydrogen peroxide which results in a mixture of VHP and water vapor. The VHP mixture is used to decontaminate isolators.

W

Warning Letter

A warning letter is a correspondence that notifies regulated industry about violations that FDA has documented during its inspections or investigations.

WEEE Directive

Waste electrical and electronic equipment directive. European Community directive 2002/96/EC where manufacturers are responsible for disposing of electrical/electronic waste.

WFI (Water for injection)

WFI is sterile and pyrogen free water containing o less than 10 CFU/100ml (Colony Forming Units) with a sample size of between 100 and 300 ml and an endotoxin level < 0.25 EU/ml.

WHO

World Health Organization.

WI

Work Instructions.

Witnessed By

When signed or initialed is legal proof that the individual signing is physically present and observes the step, calculation, or operation being performed by someone else, and that all entries of data are true and accurate.

Worst Case

A set of conditions or parameters which, in combination with product specification or attributes at their limits, pose the greatest challenge to the process.

X

--

Y

--

Z

Zone Classification

Zone classification refers to GMP areas which include controlled (aka classified) and non-controlled manufacturing areas. Areas may be classified based on EU Grades A–D and/or ISO Class 5–8 (in the US - Class 100–Class 100,000 areas.